Contradictions in Cultural-Historical Activity Theory

Joanne Hardman • Ekaterina Rzyankina

Contradictions in Cultural-Historical Activity Theory

Using CHAT to Analyse STEM Education

Springer

Joanne Hardman
School of Education
University of Cape Town
Cape Town, South Africa

Ekaterina Rzyankina
University of Cape Town
Cape Town, South Africa

ISBN 978-3-032-28649-9 ISBN 978-3-032-28650-5 (eBook)
https://doi.org/10.1007/978-3-032-28650-5

This Springer imprint is published by the registered company Springer Nature Switzerland AG
The registered company address is: Gewerbestrasse 11, 6330 Cham, Switzerland

In loving memory of my father, the legendary Peter Joseph; and of course, to my daughter Bella, who is my inspiration. Without you, there would be none of this.
JH
I would like to dedicate this work to my family—Zulfia, Sergey, Natasha, Noor, and my beloved daughter Leilah, as well as to all my friends all over the world. Thank you for your constant support and memorable moments that have enriched my life.
ER

Foreword

This is a timely and welcome book, for scholars new to cultural-historical activity theory (CHAT) and for those familiar with its concepts and methods. A world entrenched in multiple, belligerent crises needs more than through-the-window research from an objective distance. It needs research that joins struggles on the ground, takes a stance, and gets concretely involved in creating change. CHAT provides a crucial means to do so, increasingly linked to the wider scholarship of resistance (Stetsenko, 2024) and energetic interventions that challenge the inevitability of the status quo, even in the face of such giants as homelessness or even capitalism (Sannino, 2020).

One of the reasons CHAT is so powerful is due to the way it conceives and works with contradictions. They are not awkward problems to be resolved, but recognised as historically emerging driving forces for change. This view reflects the Marxist foundations of CHAT, foundations that also manifest in a rebellious, agitated gist that refuses to merely stand by and watch the world, but instead gets involved in changing it.

CHAT continues to evolve in exciting ways. Stetsenko's (2017, 2024) transformative activist stance (TAS) offers us a call to radical action based on a fully elaborated ethico-ontoepistemology. Sannino's (2015, 2023) transformative agency by double stimulation (TADS) pushes Vygotsky's understanding of how mediational means are central to volitional action, showing how transcending conflicts of motives can underpin significant change that goes against the normative grain. Voices from the Global South are stronger than ever, exploring how we might take up CHAT towards decolonising ends, bringing CHAT into dialogue with Indigenous ways of knowing (e.g., Muthivhi, 2026). Hardman's (2024) own work approaches decoloniality in pedagogy from a classical Vygotskian foundation. This subsequent book follows a similar theme of re-turning to core CHAT ideas while addressing contemporary pedagogical and social challenges.

Amid these exciting developments in CHAT, Hardman and Rzyankina offer something unusual and valuable: a whole book focused on contradictions, primarily from a CHAT perspective. This complements other recent volumes that offer comprehensive introductions to CHAT (e.g., Lémonie, 2025). For scholars newer to

CHAT, Hardman and Rzyankina's book offers an accessible way in through the foundational idea of contradictions—bringing us into contact with other ideas such as dialectics and perezhevanie. For scholars more familiar with CHAT, this book invites us to pause and dwell on contradictions, something I found to be very good intellectual hygiene as a CHAT scholar myself. They argue that analysing contradictions is inherently interpretive work. The interpretations presented in this book offer a playful mix of conventional (within CHAT at least) views traced carefully to key texts, and at times challenging and surprising takes that might invigorate our thinking.

Whether as an introductory text, a suite of empirical examples, or a way to jostle with familiar ideas, this book offers something for diverse readers, and should especially be welcomed for its unabashed positioning from the Global South.

Hardman, J. (2024). *A cultural-historical approach towards pedagogical transitions: Transitions in post-Apartheid South Africa*. Bloomsbury.

Lémonie, Y. (2025). *Transforming & understanding: An introduction to cultural-historical activity theory*. Peter Lang.

Muthivhi, A. E. (2026). *Decolonizing pedagogy in post-apartheid South Africa*. Cambridge University Press.

Sannino, A. (2015). The principle of double stimulation: a path to volitional action. *Learning, Culture and Social Interaction*, *6*, 1–15. https://doi.org/10.1016/j.lcsi.2015.01.001.

Sannino, A. (2020). Enacting the utopia of eradicating homelessness: Toward a new generation of activity-theoretical studies of learning. *Studies in Continuing Education*, *42*(2), 163–179. https://doi.org/10.1080/0158037x.2020.1725459.

Sannino, A. (2023). Toward a power-sensitive conceptualization of transformative agency. In N. Hopwood & A. Sannino (Eds.), *Agency and transformation: Motives, mediation and motion* (pp. 35–55). Cambridge University Press.

Stetsenko, A. (2017). *The transformative mind: Expanding Vygotsky's approach to development and education*. Cambridge University Press.

Stetsenko, A. (2024). Reclaiming the tools of the past for today's struggles: Radicalizing Vygotsky, via Marx, in dialogue with Audre Lorde. In A. Levant, K. Marukami, & M. McSweeney (Eds.), *Activity theory: An introduction* (pp. 369–406). ibidem.

University of Technology Sydney
Ultimo, NSW, Australia
University of Stellenbosch
Stellenbosch, South Africa

Nick Hopwood

Preface

In this book we aim to contribute to research in the field of cultural-historical activity theory (CHAT) by exploring how the concept of contradictions can be used in research on education in the fields of science, technology, engineering and mathematics (STEM), particularly in the Global South. It navigates the complex terrain of CHAT by dissecting contradictions through the perspectives of conceptual understanding, methodological approaches, and empirical case studies.

At its core, the book zooms into the conceptual realm, tracing the historical evolution of the idea of contradictions in CHAT. For researchers, understanding contradictions through CHAT requires a thorough engagement with both the historical foundations of the theory and its contemporary adaptations. Therefore, the book revisits the foundational theories of Vygotsky, Leontiev, and Engeström, offering a comparative analysis of their interpretations, and then expands upon these ideas in a way that is appropriate to contemporary contexts.

In terms of methodology, the book looks at how contradictions in activity systems can be identified and analysed. CHAT provides a robust analytical lens for examining the complexities of interaction, cognition, and change within cultural and historical contexts, and we look at new tools and methods that are designed to capture the dynamic and often elusive nature of contradictions in social and educational settings.

The empirical dimension of the book features a series of in-depth case studies in the field of education that illustrate the real-world manifestations of contradictions. These not only demonstrate how contradictions can drive change and development but also provide practical insights into managing and leveraging these contradictions for positive outcomes. With a particular focus on STEM education and the integration of digital technologies, we examine how contradictions, rather than being solely seen as disruptive, can be transformative forces that educators can harness to improve teaching and learning. Special attention is given to the tensions between traditional educational practices and the evolving demands of technology-driven learning environments, highlighting how these contradictions impact both educators and students.

While the notion of contradictions has been written about in cultural-historical activity books, we can find no book dedicated specifically to the notion of, and historical development of, contradictions. We hope that, whether you are working in education or conducting research in this field, this book provides a fresh perspective on how contradictions can serve as opportunities for growth and advancement in the classroom and beyond.

Part I: Theory and Methodology Founded on Contradictions

This part of the book focuses on how the concept of contradiction has developed over time and across cultures as a methodological tool for analysis of potential change in a system

Chapter 1 outlines the development of cultural-historical activity theory (CHAT) by looking at how it unfolded in several distinct but interconnected 'generations', each contributing to the theoretical refinement of contradictions as central to understanding change in human activity. The first generation (1920s–1930s) laid the theoretical foundations, with a focus on the relationship between individual action and societal structures. During the second generation (1950s–1960s), contradictions remained peripheral, but there was a recognition that human activity is not inherently harmonious. The third generation (1970s–1980s) marked a critical shift as contradictions became central analytical tools. The fourth phase (1990–2025), which could be considered to be unfolding now, highlights contradictions in a globalised and digital world.

Chapter 2 traces the historical and dialectical development of contradictions within Western, Soviet, and Global South philosophies, as well as looking at the way that contradictions are used within CHAT. By examining dialectics, *perezhivanie* and drama, we trace how the notion of contradiction has evolved, and how its meaning has been shaped by different intellectual contexts. We begin with a focus on Hegelian dialectical idealism before situating our work in the Marxian notion of dialectical materialism.

Chapter 3 provides a methodological lens for identifying and analysing contradictions in empirical research in education, with particular attention to contexts shaped by socio-cultural differences and economic inequalities. In this chapter, we outline various methodological techniques for using CHAT to investigate contradictions as they arise discursively. We note that contradictions can never be seen outside of their discursive manifestations, and we illustrate in this chapter how one can look for these to analyse contradictions.

Part II: Case Studies Using Contradictions in Educational Research

In Part I we explored the concept of contradiction both conceptually and from a methodological perspective. In Part II we shift focus to the practical dimensions of contradictions, drawing on illustrative cases that demonstrate how contradictions manifest in real-world activity systems. This part will examine empirical examples of contradictions as understood through the second and third generations of CHAT, highlighting how these theoretical developments can inform and guide transformative practices in educational settings.

Chapters 4 and 5 are both about CHAT research that was undertaken to address the research question: How do students and lecturers engage with engineering e-textbooks? Chapter 4 discusses the digital literacy practices involved in reading e-textbooks that surfaced in the focus-group discussions with students and in individual student interviews, while Chap. 5 presents the contradictions that surfaced in the activity of using e-textbooks. Even though these chapters use the same data, in Chap. 4 we focus on the tensions and contradictions that arise in the students' activity system, while in Chap. 5 we focus on the tensions and contradictions that arise in the lecturers' activity system.

Chapter 6 moves the discussion from tertiary education to primary school education and traces the contradictions that arise between a teacher's intended and enacted pedagogy in the presence of novel technology in computer-based mathematics lessons. Using interview data in which a deductively informed questionnaire captures data, the chapter illustrates how the introduction of new mediating means, in this case computer-based technology, can have a regressive, rather than progressive impact on teaching and learning.

Chapter 7 introduces a novel model of pedagogy for teaching from a cultural-historical foundation. This chapter doesn't utilise CHAT as such, but rather uses the work of Vygotsky with the addition of the notion of contradictions to illustrate how pedagogy can be developmental through harnessing contradictions in the form of cognitive conflict. The chapter compares the pedagogical praxis of two teachers, one who has been taught to use the novel pedagogy and one who has not been taught to use it. Findings indicate that the use of the novel pedagogy opens more dialogue in the classroom and provides for more instances of cognitive dissonance, where development can occur.

While these case studies are drawn from South Africa in the global south, they provide examples of analysis of contradictions that extends beyond the confines of this context. There is a sense, however, in which South Africa is a unique research context given its history of apartheid, which was state-sanctioned segregation and oppression on a scale not seen, in its foundational legality, elsewhere in the world. What this allows, however, given the progress from apartheid to democracy, is a view of change as it happens in real time in crisis moments. The socio-historical context of a changing South Africa provides instances of change-in-action that one may not see in many other contexts. This provides a fecund space for investigating contradictions as they arise, are navigated and potentially resolved.

Cape Town, South Africa Joanne Hardman
Cape Town, South Africa Ekaterina Rzyankina

Acknowledgements

No book is a sole project, and we would like to acknowledge the assistance we have received along this route from various people. First, we thank our participants who took part in our research reported in the book. Without willing participants there is no empirical evidence of claims made. We would like to thank our initial Springer contracting editor, Astrid Noordemeer for encouraging us to write this book. Jeanne Maclay-Meyers provided much needed critical engagement with our work, and the book is better because of her input. We have had interesting input and discussions with colleagues throughout the writing process. Warren Lilley was always there to listen to our discussions on contradictions, and we benefitted greatly from the incredibly insightful work of Andy Blunden, whose online Marxist.org is a treasure trove of all things related to contradictions in the Marxian and Hegelian frame. Our work draws heavily on the body of work from Annalisa Sannino and Yrjo Engeström, whose elaboration of the dialectical nature of contradictions in the twenty-first century has been groundbreaking in our field. Without their work, this book would have been a different project. Prof Azeem Badroodien provided much needed support for us to continue with this project, and we thank him for that. Rose-Anne Reynolds has been a constant cheerleader through this project and we thank her for her care, friendship, and laughter. Finally, but by no means last, we would like to thank our families: Ian, thank you for always thinking I could. My mom Valerie-Anne Hardman, for all the encouragement and support (JH). Bella, thanks for cooking and allowing mom (JH) to work and occasionally ignore you. I would like to thank all my family and friends, especially my beloved daughter Leighla, Noor, Zulfia, Sergey, and Natasha (EK).

Competing Interests The authors have no competing interests to declare that are relevant to the content of this manuscript.

Contents

Part I Theory and Methodology Founded on Contradictions

1 The Origins and Development of Cultural-Historical Activity Theory (CHAT) 3
Introduction 3
First-Generation Activity Theory: Vygotsky's Cultural Historical Theory 4
Mediation in the Zone of Proximal Development 6
Second-Generation Activity Theory: Aleksei Leontiev's Activity Theory 8
Individual Action Within Collective Activity 9
Third-Generation Activity Theory: Engeström and the Cultural-Historical Activity Theory (CHAT) Movement 11
Expansive Learning in the Change Laboratory 14
Fourth-Generation Activity Theory: A Work in Progress? 15
Conclusion 16
References 17

2 Conceptual Understandings of Contradiction 19
Introduction 19
The Concept of Contradiction in Philosophy 20
The Evolution of the Concept of Contradiction in Western and Soviet Philosophy 20
How the Concept of Contradiction in Cultures of the Non-Western Resonates with CHAT 26
Contradictions in Dialectics and Dialogics 29
The Role of Contradictions in Psychological Development 32
The Experiential Prisms of *Perezhivanie* 32
The Dramatic Frame of Personality 37
Contradictions Within an Activity System 38
The Nodes of an Activity System 38
Activity System Analysis 39

A Typology of Contradictions. 39
Contradictions as Developmental Mechanisms 41
Conclusion 43
References. 43

3 Contradictions as a Methodological Tool in CHAT-Based Educational Research 47
Introduction. 47
Formulating a Research Question for a Study That Will Utilise Contradictions. 48
Planning CHAT Research That Will Mobilise Contradictions 50
Gathering Data in a Way That Will Allow Contradictions to Emerge . . 52
Analysing Contradictions Within Data 55
Analysing Contradictions Using Traditional Tools. 56
Analysing Contradictions Using Data-Analysis Software 58
Reporting on Contradiction-Focused Research 64
Resolving Resolvable Contradictions (See Table 3.8) 65
Raising Unresolved Contradictions for Further Investigation 66
Conclusion 66
References. 67

Part II Case Studies Using Contradictions in Educational Research

4 When the Mediating Artefact Disrupts the Original Object: How e-Textbooks Impact Engineering Students' Learning of Physics 71
Introduction. 71
A CHAT Analysis of the Students' Activity System. 72
How Contradictions Manifest When Students Are Required to Use e-Textbooks. 73
Contradiction A: Learning Physics with e-Textbooks Versus Learning How to Operate e-Textbooks. 73
Limited Usage of e-Textbooks Versus Investing Energy to Explore the Functionalities of e-Textbooks 77
Contradiction D: Preferences for Solitary Study Versus the Collaborative Study Affordances of e-Textbooks. 79
Contradiction E: Pre-Curated Information in e-Textbooks Versus Unreliable Online Resources 80
Conclusion 81
References. 83

5 The 'Culturally More Adept Others' Who Prescribe e-Textbooks: How Are Lecturers at a University of Technology Navigating Their Evolving Role? 85
Introduction. 85
How Do Contradictions Manifest in the Lecturers' Activity System When e-Textbooks Are Used?. 86
How Do e-Textbooks Affect the Role of the Lecturer? 88

How Suitable Are Foreign e-Textbooks for the Local Context? 91
How Enabling Are e-Textbooks in the Face of Device Limitations? ... 93
How Do Contradictions Manifest Between Intersecting Systems When e-Textbooks Are Used? 95
How Will the e-Textbook Be Useful to Lecturers in Helping Them to Meet Their Responsibility of Teaching Physics? 97
How Well Does Reading the e-Textbook Fill Students' Theory Gap? 98
Conclusion 99
References 100

6 Intended Versus Enacted Pedagogy: Technology Triggering Behaviourist Pedagogy in Computer-Based Mathematics Lessons 101
Introduction 101
How Technology Can Improve Attainment in Mathematics 102
Behaviourist Software 103
Constructivist Software 103
The Study of ICT Usage in Grade 6 Mathematics Lessons at Sonskyn Primary School 106
The Context of the Study 106
The Research Design 107
The Teacher 107
The School 108
The Subject 109
The Mediating Artefact 109
Rules 110
Object 112
The Community 113
Division of Labour 114
Conclusion 118
References 118

7 A Dialectical Model of Pedagogy: Beginning Knowledge as a Site of Cognitive Development 121
Introduction 121
The Social Context 122
The Concepts 122
The Pedagogical Model: Teach-Assess-Teach 123
The Study of Two Pedagogies in Grade 2 Life Skills: Beginning Knowledge Lessons 125
The School 125
The Teachers 126
The Students 126
The Subject 126
Data Collection and Analysis 127

Results and Discussion ... 127
Conclusion ... 133
References ... 134

8 Managing the Potential Within Contradictions to Guard Against Regressive Outcomes ... 135
Introduction ... 135
A Dialectical Perspective on the Power of Contradictions in Educational Contexts ... 135
CHAT'S Thesis: Contradictions Can Drive Progressive Change ... 136
The Logical Antithesis: Contradictions Can Drive Regressive Change ... 136
The Synthesis: Contradictions Can Drive Progressive or Regressive Change ... 137
Empirical Insights from our CHAT Research on STEM Education in South Africa ... 137
Empirical Insights on Activity Systems in a University of Technology ... 138
Empirical Insights on Teaching in Primary Schools ... 138
Synthesising the Insights from Both Sets of Research ... 139
Implications for Research, Policy and Practice ... 140
Issues for Further Investigation ... 141
How the Global South Can Use the Concept of Contradiction Fruitfully for Educational Progress ... 141
Perezhivanie and the Dramatic Frame of Personality as Tools for the Global South ... 142
Conclusion ... 144
References ... 144

Appendices ... 145

References ... 147

Part I
Theory and Methodology Founded on Contradictions

Chapter 1
The Origins and Development of Cultural-Historical Activity Theory (CHAT)

Abstract This chapter traces the history of cultural-historical activity theory (CHAT) from its roots in the work of the Russian psychologist Vygotsky, through the praxis focus of Leontiev's work to Engeström's development of CHAT into its current form. The chapter focuses on the notion of mediation as the guidance of a more competent other in development and goes on to elaborate how Engeström has added as almost sociological lens to the study of psychology by locating individual actions within collective activity systems that are motivated by objects. The uniqueness of Change Laboratories as an intervention strategy that relies less on intervention than on harnessing the contradictions identified by participants in the CL closes the chapter.

Introduction

In learning theory we are faced with a conundrum: how do people learn what they do not know that they do not know?[1] Yet people do learn, and, as learning theorists, we are interested in how this happens.

In this book we look at how contradictions are dynamic sites of learning. The theory underpinning this has a long history spanning three 'generations'. It has resulted in a theoretical framework that provides a comprehensive understanding of human practices as developmental processes where the individual and social levels interact. In this chapter we trace the evolution of this theoretical framework, which offers us a mechanism for explaining how learning occurs.

[1] This is often referred to as the 'learning paradox' and is discussed by Fodor (1980)

J. Hardman, E. Rzyankina, *Contradictions in Cultural-Historical Activity Theory*, https://doi.org/10.1007/978-3-032-28650-5_1

First-Generation Activity Theory: Vygotsky's Cultural Historical Theory

Lev Vygotsky (1896–1934) was a Soviet psychologist and teacher. His focus was the mediated nature of learning, and he is best known for his sociocultural[2] theory of development. Figure 1.1 shows how Vygotsky saw a person as interacting with the world by means of cultural artefacts (although he himself did not represent his theory in such a diagram). The primary source of development lies in the use of mediational means, represented at the apex of the triangle. The subject (an individual or group) uses mediational means to act on the object of the activity. The object of the activity is the need that the subject seeks to meet. The object can be material (such as a need to create a sculpture out of a block of marble) or ideal (such as the need to teach a concept, such as photosynthesis). All activity is object-oriented because a person does not act without some reason for doing so. The natural relationships represented at the base of the triangle are subsumed under the cultural relationships represented at the apex of the triangle (Davydov, 1975, 1982). The primary mediator of human consciousness for Vygotsky is language (1986).

What one sees in Fig. 1.1 is that the subject, let us say for example a teacher, uses mediational means (mostly language but perhaps technology such as an iPad or even music) to act on the object of learning (the student's understanding of concepts). The teacher then, is the more competent other in this figure.

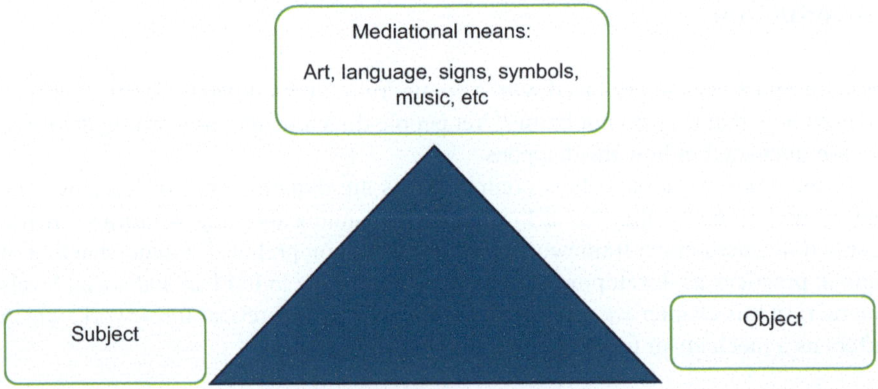

Fig. 1.1 First-generation activity theory: A graphic representation of Vygotsky's notion of mediation. Source: Hardman (2008, pg; 49). We note that Vygotsky himself does not refer to his work at all as a theory of activity. The use of this work in this diagram draws attention to the later use of Vygotsky's foundational work in the underpinnings of both Leontiev's activity theory and Engström's CHAT

[2] It should be noted that the terms cultural-historical and sociocultural are used interchangeably in literature, with the Western literature favouring the name sociocultural and the Russian literature focusing on cultural-historical when naming the theory.

Vygotsky postulated that all knowledge, all higher cognitive functions unique to human thinking, begin as actual relationships outside of the developing child (1986). In his general genetic[3] law, Vygotsky indicates that what initially begins as meaning generated between a culturally more adept other and a child then becomes internalised as the child's own knowledge, through development.

This dialectical view of development lends itself to a view of teaching as leading development, where children or novices come to learn meaningful knowledge through interacting with those in their context who are more knowledgeable, at this point, than they are. According to Vygotsky, the world is never approached directly during the development of higher cognitive functions but is always mediated (Bateson, 1972; Wertsch, 1991, 1998). To study development Vygotsky utilises a method he refers to as a microgenetic method, where 'genetic' refers to the genesis of what is observed. Vygotsky's microgenetic method represents a developmental research approach that examines psychological processes as they unfold in real time, capturing the dynamic transformation of cognitive functions during their actual formation. This methodological framework is predicated on the principle that psychological phenomena are most fully understood when observed in their process of change rather than as static, finished products.

The microgenetic method involves the systematic observation and analysis of psychological development compressed within relatively brief timescales, ranging from minutes to weeks, in which researchers create conditions to provoke or accelerate developmental changes. By focusing on documenting moment to moment cognitive activity, the researcher can trace qualitative shifts and reorganisation of mental functions as they emerge and begin to stabilise.

Central to this methodology is Vygotsky's conception that higher psychological functions arise through social interaction and cultural mediation. The microgenetic method therefore focuses on capturing the process of internalisation whereby interpsychological (social, external) activity becomes intrapsychological (individual, internal) functioning. Researchers employing this method attend specifically to the role of semiotic mediation, especially language and symbolic tools, in restructuring cognitive processes.

Methodologically, microgenetic analysis requires dense, longitudinal observation during periods of active transition, with careful attention to both quantitative performance changes and qualitative shifts in the structure and organization of psychological activity. The strength of this approach lies not solely in uncovering that development occurs but the process by which it does so, illuminating the mechanisms, mediational means, and dialectical processes through which nascent cognitive capabilities are developed and consolidated.

---

[3] The term 'genetic' means 'related to the beginning' and does not refer to biological genes.

Mediation in the Zone of Proximal Development

For Vygotsky, good learning, that which leads to development, begins between the mediator and the novice in a specific developmental zone, which he called, in Russian, the *zona blizhayshego razvitiya*. In English this zone is referred to as the 'zone of proximal development' (ZPD), although Smagorinsky (2018, pg. 254) has proposed that it would be better translated as the 'zone of next development'. It is in this space that the culturally more competent 'other' (a teacher in a class, for example) guides the child beyond what they are currently able to accomplish on their own.

While the more expert other certainly has more knowledge than the novice, the development of the novice's knowledge requires a co-construction of meaning between the novice and the expert. An inability to co-construct meaning leads to the novice acquiring mere definitions of concepts, rather than 'owning' them by relating them to their own understanding of the world.

This differs from a view of learning that promotes a more individualist explanation of learning, such as the genetic epistemology[4] of the Swiss psychologist Jean Piaget (1977) (who is discussed further in Chap. 6). For Vygotsky, mediation was the key to understanding how children learn, and it provided the basis for developing a theory of education.

A fundamental premise of Vygotskian theory is that basic biological processes are transformed into higher cognitive functions using culturally meaningful tools and signs (such as language) during social interaction (Vygotsky, 1978). That is, children are born with certain basic, biological processes, such as perception and the potential ability to recall an image from memory (Diaz et al., 1990). As the child develops within the social world, these elementary processes are transformed by the child's interaction with the social world.

Higher cognitive functions develop first as interpsychological functions, with the primary caregiver initially guiding the child's activity, and later 'turn inward' becoming intrapsychological functions. Thus it is the primary caregiver, most often the mother, who originally mediates the child's activity and externally regulates the child's interaction with the environment (Greenberg & Moll, 1990), but what begins as external regulation or social regulation turns inwards and becomes self-regulation. Higher cognitive functions, then, have social origins.

This conceptualisation challenges the dualist stance to development, where an individual's mind is seen as being distinct from society. Instead it proposes that the mind is social, which Vygostsky (1978) makes clear in his general genetic law of cultural development:

> Every function in the child's cultural development appears twice: first, on the social level, and later on the individual level; first, between people (interpsychological), and then inside the child (intrapsychological). This applies equally to voluntary attention, to logical mem-

[4] However, readers familiar with Piaget's work will find resonances in his theorisation of assimilation and accommodation in Vygotsky's work.

> ory, and to the formulation of concepts. All the higher functions originate as actual relations between human individuals. (p. 57)

From this it follows that the nature and the quality of mediation are crucial in the development of higher cognitive functioning, as well as in self-regulation.[5]

For Vygotsky (1978, 1986) the ZPD represents the gap between what a student can accomplish with assistance and what that student can accomplish on their own. This implies that a teacher, or at the very least a culturally more advanced peer, is developmentally necessary, which indicates the dialogical nature of learning (Tharp & Gallimore, 1988; Diaz et al., 1990). The concept of mediation casts pedagogy as guided assistance, where a more competent peer or teacher guides a student into established ways of knowing and being. However, as we shall see in the quote below, this is not a space of training technical skills; the ZPD is a developmental principle that refers specifically to the development of higher cognitive functions, which we would today call executive functions.

Vygotsky's concept of mediation is often confused with the concept of scaffolding, which was later developed by Wood et al. (1976) in the United States, and which refers to assisting students to develop specific skills in order so that they can complete specific tasks. However, scaffolding is not developmental (Hardman, 2023), while mediation is. For example, a teacher can scaffold the alphabet to students in Grade 1 by teaching them rhymes or songs, but this does not mean that the child has developed the ability to read. While learning the alphabet can be done by rote with little understanding of what the discrete letter means or could mean, learning to read takes place over time, it is a developmental task that takes on average 3 years to accomplish in order to read not just words, but to read for understanding. This cannot be scaffolded in a single lesson but must take place over time.

For Vygotsky, education was developmental rather than instructional. This is apparent when Vygostsky states (1987):

> To implant [something] in the child… is impossible… it is only possible to train him for some external activity like, for example, writing on a typewriter. To create the zone of proximal development, that is to engender a series of processes of internal development we need the correctly constructed processes of school teaching. (p.134)

The ZPD represents a move in Vygotskian theory from focusing on sign-mediated actions to socially mediated actions (Greenberg & Moll, 1990), and this highlights the importance of educational intervention (Hedegaard, 2001).

What one can see from this quote is that Vygotsky's (1978) work is a conscious attempt to develop a Marxian psychology, highlighting that the consciousness of humans is inextricably related to the material and social world. Drawing on Marx's concept of labour and tool use as uniquely human activities that transform both the external world and human consciousness, Vygotsky proposed that psychological tools (language, signs, symbols) mediate cognitive development just as physical

[5] Recent developments in neuroscience have proven that executive functioning, those thought processes that make us uniquely human, arise as products of our interaction with the social realm.

tools mediate material production. These cultural tools are historically developed and socially transmitted. Employing Marx's dialectical materialism, Vygotsky illustrated how change occurs through the resolution of tensions[6] between the individual and society, the biological and the cultural, leading to qualitatively new forms of consciousness. Of course, Vygotsky did not work alone; this would fly in the face of all he believed about development. One of his most important collaborators on cultural-historical theory was Alexander Luria, whose work with Vygotsky in Uzbekistan is a formative example of how cognition develops (Luria, 1976). This work was only published in 1976 in the West but continues to inform theorists who use cultural-historical theory as a foundation for understanding development.

Second-Generation Activity Theory: Aleksei Leontiev's Activity Theory

Aleksei Leontiev (1903–1979) was a colleague of Vygostsky's in Moscow who further developed Vygotsky's ideas. The object of activity for Leontiev, could be 'either material or ideal, either present in perception or existing only in the imagination or thought' (1978, p. 62). This is similar to how Vygotsky viewed the object of activity, noting though, that the difference lies in their focus on the primacy of language as a mediating means and the collective nature of an object for Leontiev. For him the object is the motivation of a collective activity while an action is goal oriented. Rather than emphasising language, Leontiev focused on how practical labour activity mediates cognitive development (Kozulin, 2003).[7]

While Vygotsky's theory points the way towards an understanding of learning as distributed, it does not provide an analytical framework capable of situating learning within a wider context, accounting for the collective and dynamic nature of activities (Engeström, 1987). In addition, the distinction between individual and group acts is implied, but not articulated in Vygotsky's theory. Leontiev explored this explicitly, and, in this way, he elaborated on Vygotsky's ideas to develop the analytical framework of activity theory.

[6] It is important to note, however, the in his own work Vygotsky does not refer to contradictions explicitly. However, the dialectical nature of his work certain implies a recognition of contradictions as sites of change.

[7] Kozulin (1986) makes much of Leontiev's apparent rejection of the developmental importance of semiotic mediation; there is however, sufficient evidence from Leontiev's writing to indicate that he did recognise the importance of semiotic mediation, but focused his attention more on practical rather than semiotic activity (see Engeström, 1987, for a discussion here, and Daniels, 2001 for an interesting insight into the historical conditions that militated against Leontiev focusing on semiotic mediation).

Individual Action Within Collective Activity

Leontiev drew a distinction between individual action and collective activity (1981), and viewed all activity as necessarily collective and involving division of labour. He illustrated that focusing on only a single action in an activity can tell us little about the actual overall activity; to understand any specific actor's actions requires that we understand the collective activity.

Leontiev's hierarchical model of functioning conceives of collective activity as driven by the object, which is social; while individual actions are directed at goals, which are conscious; while operations, which are automatic, are called into play by the tools and conditions of the action being carried out (Leontiev, 1981; Engeström, 1987). This three-level model of activity is represented in Fig. 1.2.

In this formulation, Leontiev illustrated how motives, emotions and creativity are social endeavours, which is difficult to do with Vygotsky's model. This is achieved in Leontiev's work because his model of activity situates individual, goal-directed actions in the social context of an activity. Take, for example, emotions, such as happiness or distress. These are meaningless outside of a social realm; an infant's tears have meaning only in a social context and achieving the goal (being fed, perhaps) is a social rather than individual endeavour, in the sense that it requires social co-operation.

In Leontiev's work we find a sense of how individual actions play out against the meaningful background of a social activity. There is a certain nuance here between an individual using action to meet a goal, which may differ from the overall object of the activity he/she is engaged in. This distinction between object and goal/individual and collective underpins Engström's later development of this theory. This is

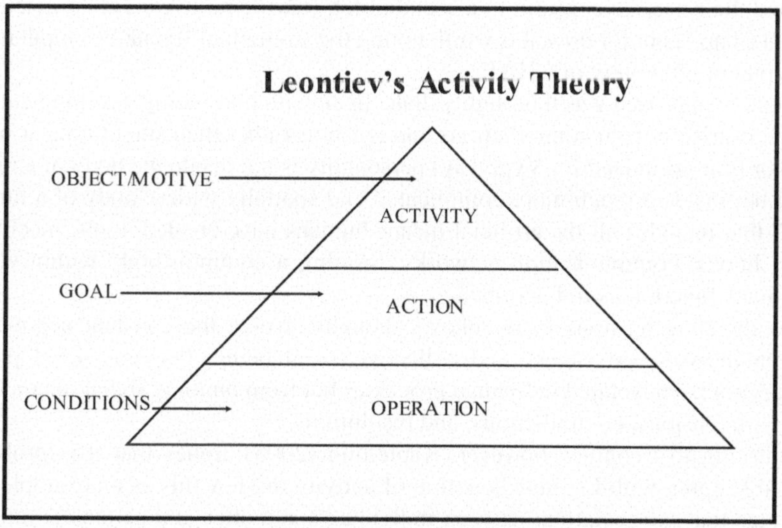

Fig. 1.2 Second-generation activity theory. Source: Daniels (2001, p. 87)

something that is certainly hinted at in Vygotsky's work, but is ultimately not theorised. Leontiev enables us to see activity as a social endeavour, whereas Vygotsky's work is still located at the level of the individual acting with mediational means, albeit in a social realm where the object is never approached without mediation.

The hierarchical nature of activity is illustrated elegantly in Leontiev's example of the primeval hunt. Here Leontiev (1981) demonstrates the weaknesses inherent in focusing solely on individual actions outside of the activity against which they become meaningful:

> When a member of a group performs his labour activity he also does it to satisfy one of his needs. A beater, for example, taking part in a primeval collective hunt, was stimulated by a need for food or, perhaps, by a need for clothing, which the skin of the dead animal would meet for him. At what, however, was his activity directly aimed? It may have been directed, for example, at frightening a herd of animals and sending them toward other hunters, hiding in ambush. That, properly speaking, is what should be the result of the activity of this man. And the activity of this individual member of the hunt ends with that. The rest is completed by the other members. This result, i.e., the frightening of game, etc., understandably does not in itself, and may not, lead to satisfaction of the beater's need for food, or the skin of the animal. What the processes of his activity were directed to did not, consequently, coincide with what stimulated them, i.e., did not coincide with the motive of his activity; the two were divided from one another in this instance. Processes, the object and motive of which do not coincide with one another, we shall call 'actions'. We can say, for example, that the beater's activity is the hunt, and the frightening of game his action. (p. 210)

This example of the primeval hunt has been used by theorists such as Engeström (1987) to suggest that Leontiev draws a distinction between individual actions and collective activities. The distinction between individual action and collective activities is taken up in the work of Evald Ilyenkov, also known as a second generation activity theorist. Ilyenkov (1977), uses the notion of dynamic tensions to illustrate how development occurs through a continuous negotiation between the individual's goal and their need to conform to a social order, sharing motives. This is discussed more in Chap. 2 but for now, it is worth noting the strength of Ilyenkov's input to the development ultimately of CHAT.

Ilyenkov approach was thoroughly dialectical, with him seeing development and change as arising from contradictions and tensions rather than smooth linear progress. For him (as indeed for Vygotksy) personality is not inside the body of an individual but inside a much more complicated and spatially wider "body of a human being" that includes all the artificial means humans have created, tools, machines, words, books, communication networks, forming a common body within which individuals function as living organs.

The dynamic tensions in Ilyenkov's thought involve the constant negotiation between individual existence and collective/social being. Personal development emerges not from isolated individual processes but from ongoing social interactions filled with tensions, contradictions, and resolutions.

Returning to Leontiev, however, Kaptelinin (2005) argues that it is probably more in keeping with Leontiev's notion of activity to view this as an example that illustrates that 'dissociation between individual's activities and actions… initially emerges as a result of division of labour in collective activities'[8] (p. 12). This dissociation between individual actions and collective activity carries with it echoes of the alienation that an individual has from the product of their work, in a Marxian sense.

Further, by indicating how division of labour is historically implicated in the development of higher cognitive functions, Leontiev points to the hierarchical structure of activity implied by division of labour. However, outside of his discussion of the hunt, Leontiev does not theorise how differential division of labour influences the individual within an activity. He does not elaborate the social outside of the notion of the activity as collectively object-oriented, leaving notions of social rules under-theorised. He does not discuss how social rules afford and constrain individual actions or how they dynamically impact on an activity. Ilyenkov's work, too, fails to situate actions and activities within a rule bound context. This is where Engström's (1987) work adds depth to understanding how collective activity functions in an activity system.

Third-Generation Activity Theory:[9] Engeström and the Cultural-Historical Activity Theory (CHAT) Movement

The Finnish academic Yrjö Engeström (1948–) is the main figure in the third generation of activity theory. However, after the work of Vygotsky and Leontiev was translated into English, there have been other academics who have built on their ideas. These include the American psychologist Michael Cole, who, in 1996, was the first to use the acronym 'CHAT' for 'cultural-historical activity theory', as well as the Danish psychologist Mariane Hedegaard and Anne Edwards who represent the developmental, psychological as well as pedagogical aspects of the movement respectively, and the significant input of Anna Stetsenko, with her focus on a transformative activist stance and Annalisa Sannino, whose work in change laboratories continues to grow the field. In all these different iterations of the development of Vygotsky's initial work, there is a focus on how humans develop in the social world. Hedegaard's (2001) work will be referred to in Chap. 7 of the current book. Here we merely mention that her pedagogical elaboration of the double-move in pedagogy,

[8] I recognise that the primeval hunt can indeed be interpreted as Engeström does, in other words, as an indication that actions are individual while activities are collective. However, on my reading of Leontiev's (1981) work I am inclined to agree with Kaptelinin that his psychological bias seems always to point to the individual. Activities are carried out by individuals, either individually or collectively, who are animated in the activity by the object (the motive that drives them to act). For a well argued but different view, I refer readers to Engeström (1987).

[9] Note, however, that some authors refer to this as second generation activity theory (see, for example, Daniels, 2001 and Roth & Lee, 2007). In this book, however, we maintain the distinction between Leontiev and Engeström's conceptualisation of human activity, and therefore refer here to Engeström's work as third generation activity theory.

as that space in which the ZPD is opened and local and abstract knowledge become intertwined, has had a profound impact on our own work with pedagogy as a developmental tool. For the purposes if this chapter, however, our focus narrows to the work of Engström (1987), which can be seen as ground breaking in relation to mapping human activity.

Engeström builds on the theories of Vygotsky and Leontiev and offers the framework of an activity system, which shows how individual actions are transformed into shared, collective objects through interactions with community members, and how division of labour impacts on individual actions in a collective activity. Figure 1.3 represents Engeström's third-generation[10] CHAT heuristic (Spinuzzi, 2020.

Figure 1.3 shows that the subject (which could be singular or plural) acts on the object to transform it using mediating artefacts in order to arrive at specific outcomes. In turn, the subject's position is influenced by the rules of the system, the community and the division of labour (how the context is organised, which refers also to vertical and horizontal division of labour) (Engeström, 1987, 1999; Daniels, 2001).

Engeström provides an example of a primary-care unit activity system that has as its subject a medical doctor, where the subject is also a member of the community who share a common object. Here the object is the patient's health and various artefacts are used to act on this object (such as a stethoscope and medicine) in a context where the doctor and nurses enact different roles, tasks and responsibilities as part of the medical community. Various rules such as what time the practice opens and how many patients are seen a day influences the functioning of the activity system.

This allows for a nuanced view of human activity as dialectical, both social and individual, with neither individual nor social serving as a starting point for

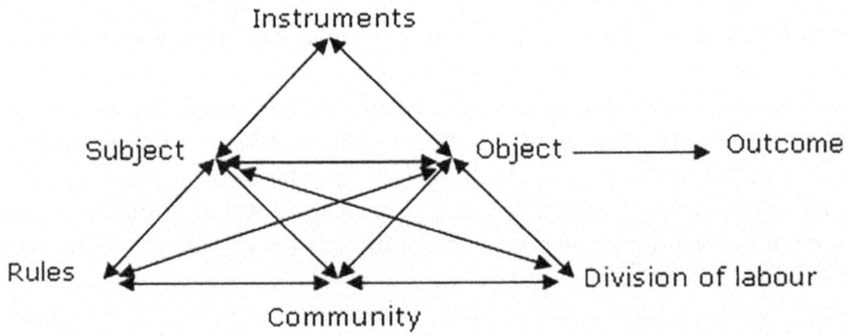

Fig. 1.3 Third-generation activity theory: Engeström's activity system. (Source: Engeström, 1987, p. 78)

[10] We are aware that this is generally referred to in the literature as second generation activity theory which we are cognisant of (Engeström, 1987; Engeström, 1999; Spinuzzi, 2020). We feel that this tends to historically silence the input of Leontiev's work which is subsumed in this activity triangle. We argue that the significant input of Leontiev's focus on collective activity should stand alone as a generation of activity theory in its own right.

explaining the other (Roth & Lee, 2007). The notion of an activity system, then, illustrates how one might understand Leontiev's suggestion that actions can only be understood against the background of an activity; here we have a theoretical idea of what that 'activity' incorporates.

Arising from Engeström's doctoral work (1987) this contemporary and popular version of CHAT is premised on the notion of learning as 'expansive'. The bi-directional arrows indicate the dynamic nature of the nodes of the triangle. In this expanded version of Vygotsky's triadic formulation of mediated action, the context of power relations and rules influences the subject's actions. According to Engeström, there are three mediating relationships within this triangular representation: mediating artefacts (tools and signs) mediate between the subject and the object; rules mediate between the subject and the community; and division of labour mediates between the community and the object. These mediating relationships need to be seen in terms of the larger context of the triangle.

As CHAT has developed in the twenty-first century, Engström's work has conceived of activity systems as multiple rather than singular entities. While a person may act within one activity system, they are part of many other systems that afford and constrain how they act. Figure 1.4 provides a more updated version of CHAT with its focus on multiple activity systems interacting with each other at any one time.

In Fig. 1.4 we can see two activity systems acting together, as they so often do. For example, an organisation and a household interact when a person plays in a role in both systems.

To further illustrate this I will point out that I am a lecturer at a university and a mother in a household. I still inhabit both systems when I am at the university, although I am obviously more focused on my teaching in this context. But if I receive a call from my child's school, saying she is ill and I must fetch her, I set my

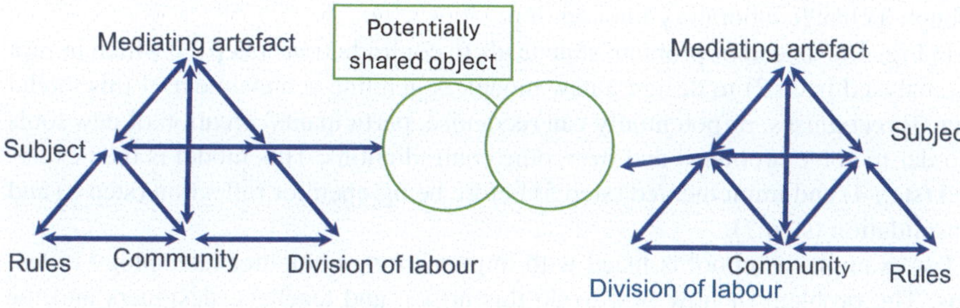

Fig. 1.4 An extension of third-generation activity theory to encompass intersecting systems. Source: Engestrom, 1987; pg 56

class a task to do while I am away. Anyone watching the students doing this task without knowing I inhabit another activity system will assume my pedagogical practice includes hands-on-tasks throughout a lecture, which it generally does not. This shows that to study anyone we need to study all the systems that they inhabit.

Expansive Learning in the Change Laboratory

In an extension of Vygotsky's work, Engeström (1987) indicated how Vygotsky's microgenetic method explains variation but is unable to explain creative activity. That is, this method is unable to account for how new activity systems come into being. Engeström (1987, 2005) met this methodological need by developing a method capable of tracking the expansive learning that people and systems undergo through interaction.

Engeström came up with the Developmental Work Research (DWR) methodology, which uses contradiction analysis to facilitate collective reflection and systemic transformation. He built on this further by developing the Change Laboratory method (under the broad umbrella of DWR), where an intervention is structured in such a way that the members of the system effect change in the system through an expansive learning cycle. This approach helps participants recognise contradictions in their own activity systems and collaboratively design solutions that address underlying systemic tensions rather than merely treating symptoms (see for example the work of Kajamaa et al., 2023, for example of a change laboratory in a Finnish school and Hopwood's 2025 review of change laboratories in schools over 25 years). One of the real strengths of Change Laboratories is the use of teacher and student voices, rather than the imposition of an intervention that doesn't take account of these lived experiences.

Figure 1.5 illustrates how a community engages in an expansive learning cycle through a change laboratory intervention.

In Fig. 1.5, a need or problem state leads to contradictions (step 1), which in turn are analysed (step 2) to design a new model. A genuine achievement of this model (step 3) recognises, or potentially can recognise, participants' creation of new tools in order to solve problems and overcome contradictions. This model is then examined (step 4) and implemented (step 5) before being open for reflection (step 6) and consolidation (step 7).

For example, a school is faced with implementing computer technology in lessons. The problem of how best to do this arises, and teachers, designers and the principal discuss what challenges they might face. Is there, for example, a possible contradiction between the use of this new technology and established pedagogy? Or perhaps a contradiction that arises between the teacher's need to exercise control over curriculum coverage and the students' desire to explore using the computers? These contradictions are discussed and analysed, and a model for implementation is agreed upon. The new model of computer implementation is then piloted in one class or with one subject. Problems with the model are discussed and worked out

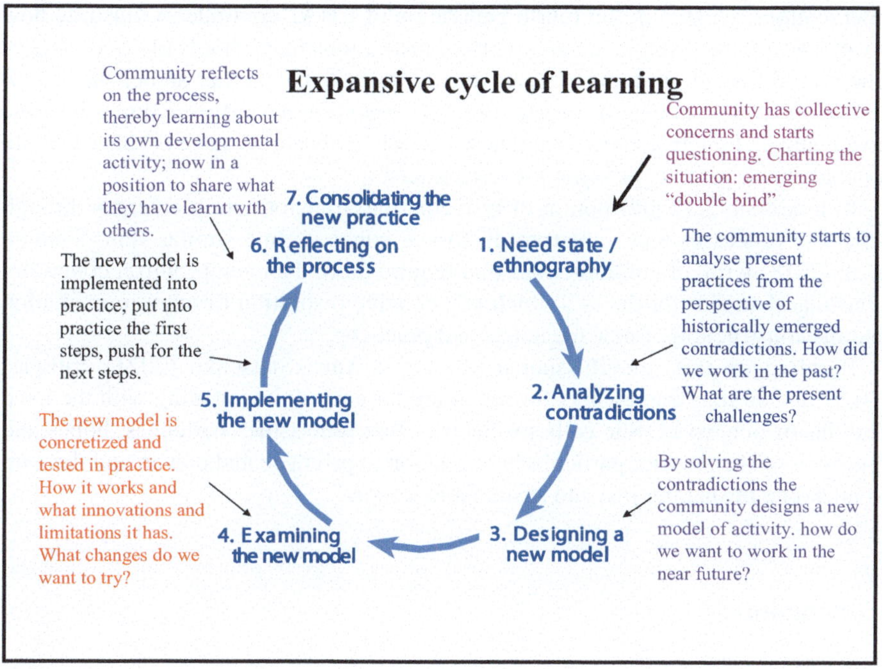

Fig. 1.5 An expansive learning cycle. Source: Adapted from Virkkunen and Newnham (2013; pg. 69)

before it is implemented within the whole school. Teachers are then called on to reflect on the process, changing it to meet their pedagogical needs.

What we have here, then, is an interventionist methodology that has some of the hallmarks of action research; a systematic, participatory approach to inquiry in which practitioners investigate and improve their own professional practice through iterative cycles of action and reflection. Where it differs from action research is in its firm theoretical base in a specific theory. This is an extremely effective methodology for studying novel learning or expansive learning in complex working spaces (Daniels, 2005, 2006).

Fourth-Generation Activity Theory: A Work in Progress?

CHAT is now evolving through its application to diverse domains, including education, healthcare and organisational change. Researchers are also integrating CHAT with other theoretical approaches, such as systems theory and complexity theory, enabling a more holistic understanding of contradictions. Attention is also being paid to new contradictions emerging from globalisation and technological change, which demand continued theoretical development. Change laboratory interventions, located in the fourth generation of CHAT continue to illustrate how change is possible even in apparently rigid institutions like schools (de Beer, 2022). The focus, now, on multiple systems interacting with each other and changes happening in various nodes of a system or between systems and over time, illustrate how complex human activity is and how the only constant we have is change. Truly a paradoxical, yet real, position for researchers.

In Lektorsky's (2009) notion of re-mediation, transformation emerges through reflective processes that reinterpret historical contexts. In a similar vein, Sannino et al. (2018) write of contradictions that require more than compromise, contradictions that demand 'thirdness', which is a creative resolution through re-mediation and the construction of new meanings and practices.

Building on this, the Russian psychologist Anna Stetsenko (2021) critiques CHAT and similar frameworks for not going far enough in engaging with the lived realities of our world. She calls for theories that reflect the challenges, hopes and injustices of the present, particularly in relation to privilege and power, and that aim to intervene in shaping just and sustainable futures.

Conclusion

Throughout its evolution, CHAT has moved from a focus on individual cognition in social contexts to a broader, systemic analysis of collective activity and societal change. Contradictions, once marginal, have become central to understanding how human practices evolve, adapt and transform.

Engeström's conceptualisation of an activity system as the basic unit of analysis (Engeström, 1987) serves as a useful tool for situating pedagogy in context. Although, for Engeström (1987), an activity unfolds over time and is not generally equated with something as brief as a school lesson (Roth & Lee, 2007), it is a fruitful way of studying practice in schools and universities, particularly when combined with Vygotsky's ideas on the importance of language in development as well as the development of a focus on activity introduced by Leontiev and developed further by Engström.

CHAT, as described by Engeström, recognises that the contradictions that arise in and between activity systems are sites of dynamic change. By positioning contradictions as productive forces rather than problems to be eliminated, CHAT provides valuable tools for analysing and facilitating transformation in educational contexts. It is this notion of contradictions as methodological tools for studying change in education that is the premise of this book.

References

Bateson, G. (1972). The logical categories of learning and communication. *Steps to an Ecology of Mind*, 279–308.

Cole, M. (1996). *Cultural psychology: A once and future discipline*. Harvard University Press.

Daniels, H. (2001). *Vygotsky and pedagogy*. Routledge Falmer.

Daniels, H. (2005). Vygotsky and educational psychology: Some preliminary remarks. *Educational and Child Psychology, 22*(1), 6–17.

Daniels, H. (2006). The 'social' in post-Vygotskian theory. *Theory & Psychology, 16*(1), 37–49. https://doi.org/10.1177/0959354306060107

Davydov, V. V. (1975). *Soviet studies in the psychology of learning and teaching mathematics* (Vol. 7). University of Chicago Press.

Davydov, V. V. (1982). *Soviet studies in the psychology of learning and teaching mathematics* (Vol. 7). University of Chicago Press.

de Beer, J. (2022). Looking at teacher professional development through a fourth-generation cultural-historical activity theory lens: The value of "change laboratories". In *Future-proofing teacher education* (pp. 112–129). Routledge.

Diaz, R. M., Neal, C. J., & Amaya-Williams, M. (1990). Vygotsky and education: The social origins of self-regulation. In L. C. Moll (Ed.), *Vygotsky and education: Instructional implications and applications of sociohistorical psychology* (pp. 127–154). Cambridge University Press.

Engeström, Y. (1987). *Learning by expanding: An activity-theoretical approach to developmental research*. Orienta-Konsultit.

Engeström, Y. (1999). Activity theory and individual and social transformation. In Y. Engeström, R. Miettinen, & R. L. Punamäki (Eds.), *Perspectives on activity theory* (pp. 19–38). Cambridge University Press.

Engeström, Y. (2005). *Knotworking to create collaborative intentionality capital in fluid organizational fields*.

Fodor, J. A. (1980). Methodological solipsism considered as a research strategy in cognitive psychology. *Behavioural and Brain Sciences, 3*(1), 63–73. https://doi.org/10.1017/S0140525X00001515

Greenberg, J., & Moll, L. C. (1990). Creating zones of possibilities: Combining social contexts for instruction. In L. C. Moll (Ed.), *Vygotsky and education: Instructional implications and applications of sociohistorical psychology* (pp. 319–348). Cambridge University.

Hardman, J. (2008). Researching pedagogy: An activity theory approach. *Journal of Education, 45*(1), 65–95.

Hardman, J. (2023). *A cultural-historical approach towards pedagogical transitions: Transitions in post-apartheid South Africa*. Bloomsbury Academic.

Hedegaard, M. (Ed.). (2001). *Learning in classrooms*. Aarhus University Press.

Hopwood, N. (2025). Twenty-five years of change laboratories in schools: A critical and formative review. *Educational Action Research, 33*(4), 571–598.

Ilyenkov, E. V. (1977). *Dialectical logic: Essays on its history and theory* (H. C. Creighton, Trans.). Progress Publishers. (Original work published 1974).

Kajamaa, A., Kumpulainen, K., & Rajala, A. (2023). Finding life beyond the classroom walls: A change laboratory supporting expansive de-encapsulation of school. *Éducation et Didactique, 17*(2), 125–141. https://doi.org/10.4000/educationdidactique.11773

Kaptelinin, V. (2005). The object of activity: Making sense of the sense-maker. *Mind, Culture, and Activity, 12*(1), 4–18. https://doi.org/10.1207/s15327884mca1201_2

Kozulin, A. (1986). The concept of activity in soviet psychology: Vygotsky, his disciples and critics. *American Psychologist, 41*(3), 264–274. https://doi.org/10.1037/0003-066X.41.3.264

Kozulin, A. (2003). Psychological tools and mediated learning. In A. Kozulin, B. Gindis, V. S. Ageyev, & S. M. Miller (Eds.), *Vygotsky's educational theory in cultural context* (pp. 15–38). Cambridge University Press.

Lektorsky, V. A. (2009). Mediation as a means of collective activity. *Learning and Expanding with Activity Theory*, 75–87.

Leontiev, A. N. (1981). *Problems of the development of the mind*. Moscow: Progress. (Original work published 1959).
Luria, A. R. (1976). *Cognitive development: Its cultural and social foundations*. Harvard University Press.
Piaget, J. (1977). *The development of thought: Equilibration of cognitive structures*. (Trans A. Rosin). Viking.
Roth, W.-M., & Lee, Y.-J. (2007). "Vygotsky's neglected legacy": Cultural-historical activity theory. *Review of Educational Research, 77*(2), 186–232. https://doi.org/10.3102/0034654306298273
Sannino, A., Engeström, Y., & Lemos, M. (2018). Formative interventions for expansive learning and transformative agency. In *Cultural-historical activity theory approaches to design-based research* (pp. 113–147). Routledge.
Smagorinsky, P. (2018). Is instructional scaffolding actually Vygotskian, and why should it matter to literacy teachers? *Journal of Adolescent & Adult Literacy, 62*(2), 253–257. https://doi.org/10.1002/jaal.756
Spinuzzi, C. (2020). "Trying to predict the future": Third-generation activity theory's codesign orientation. *Mind, Culture, and Activity, 27*(1), 4–18.
Stetsenko, A. (2021). Scholarship in the context of a historic socioeconomic and political turmoil: Reassessing and taking stock of CHAT. Commentary on Y. Engeström and A. Sannino "from mediated actions to heterogenous coalitions: Four generations of activity-theoretical studies of work and learning". *Mind, Culture, and Activity, 28*(1), 32–43.
Tharp, R. G., & Gallimore, R. (1988). *Rousing minds to life: Teaching, learning, and schooling in social context*. Cambridge University Press.
Virkkunen, J., & Newnham, D. S. (2013). *The change laboratory: A tool for collaborative development of work and education*. Sense Publishers.
Vygotsky, L. S. (1978). *Mind in society: The development of higher psychological processes*. Harvard University Press.
Vygotsky, L. S. (1986). *Thought and language* (A. Kozulin, Trans.). MIT Press. (Original work published 1934).
Vygotsky, L. S. (1987). *The collected works of LS Vygotsky: The fundamentals of defectology* (Vol. 2). Springer Science & Business Media.
Wertsch, J. V. (1991). *Voices of the mind: Sociocultural approach to mediated action*. Harvard University Press.
Wertsch, J. V. (1998). *Mind as action*. Oxford University Press.
Wood, D., Bruner, J. S., & Ross, G. (1976). The role of tutoring in problem solving. *Journal of Child Psychology and Psychiatry, 17*(2), 89–100. https://doi.org/10.1111/j.1469-7610.1976.tb00381.x

Chapter 2
Conceptual Understandings of Contradiction

Abstract Chapter 2 examines the conceptual understandings of contradiction as a core construct underpinning Cultural-Historical Activity Theory (CHAT). It begins with a philosophical genealogy of the concept, tracing how contradiction has evolved within Western and Soviet traditions. Particular attention is paid to dialectical materialism in the USSR, where contradiction became a foundational mechanism for understanding development, motion, and transformation in nature and society. The chapter then broadens the discourse to include cultures of the Global South, illustrating how indigenous epistemologies also engage with tensions, complementarities and cyclical unity, thereby resonating with CHAT's developmental orientation. From this foundation, the chapter explores contradictions within dialectics and dialogics, linking them to psychological development through the experiential prism of perezhivanie and the dramatic frame of personality. These perspectives foreground how contradiction is lived, experienced and negotiated at the level of the subject. The chapter subsequently shifts to activity-theoretical perspectives, unpacking contradiction within and between the nodes of an activity system, and outlining methodological approaches to activity system analysis. It concludes by positioning contradictions as generative and developmental mechanisms rather than obstacles, showing how they drive transformation across individual, collective, institutional and sociocultural planes.

Introduction

Contradictions in CHAT are understood as misalignments or conflicts within and between the nodes of an activity system. As CHAT has evolved, it has increasingly focused on the concept of contradictions as vital to understanding and driving change. However, the concept of contradictions within CHAT is complex, partly because of the linguistic and cultural origins of this movement.

Originating in the Soviet Union, CHAT's foundational works, including those by Vygotsky (born in Jewish family) and Leontiev (Christian family), were primarily in Russian. As a first-language Russian speaker drawing directly from Russian texts,

J. Hardman, E. Rzyankina, *Contradictions in Cultural-Historical Activity Theory*, https://doi.org/10.1007/978-3-032-28650-5_2

I hope to give a faithful interpretation of the ideas of these early CHAT theorists, while also acknowledging that they were influenced by Western philosophy and by religious traditions with their roots in the Middle East. At the same time, I want to respect the diversity in scholarly interpretations of contradictions, and the varied applications across different cultures and academic traditions. This diversity enriches CHAT and creates a landscape where contradictions are understood and applied in manifold ways.

The intention of this chapter is to describe how the concept of contradictions has evolved and to clarify conceptual ambiguities, in order to benefit those who use the concept of contradictions in their work.

The Concept of Contradiction in Philosophy

The concept of contradiction has a long philosophical lineage, and human cultures have shown a sustained fascination with the way paradoxes enable self-reflection and deeper insight. A paradox can be defined as a situation, statement or phenomenon in which two seemingly incompatible or mutually exclusive elements coexist, revealing limits in conventional logic and inviting reconsideration of what is taken to be true. Contradictions and their possible syntheses appear across both Western and non-Western traditions, from the dialectical method of Heraclitus and later Hegel, to the Buddhist and Daoist embrace of unity-in-opposites, to Indigenous and postcolonial perspectives that emphasise relationality and dynamic tension. Within these traditions, contradiction is not simply a logical inconsistency to be eliminated, but a generative force that provokes inquiry, transformation and, at times, reconciliation.

The Evolution of the Concept of Contradiction in Western and Soviet Philosophy

The concept of contradiction has ancient roots and has taken on different meanings as it has moved through the history of Western philosophy.[1] It is worth noting the significant stages in the development of this complex concept.

[1] The term *Western philosophy* is used here simply as a historical–geographical label referring to traditions rooted in ancient Greek thought and developed through European and North American intellectual history. It is not intended to imply superiority or homogeneity. Soviet philosophy, while partly drawing on Western figures such as Hegel and Marx, developed under distinct political and institutional conditions and is therefore treated as a related but separate tradition. During the Cold War, this distinction aligned with the geopolitical categories of the *First World* (capitalist industrialised countries), the *Second World* (socialist countries led by the USSR), and the *Third World* (non-aligned and post-colonial nations). Although these terms have largely fallen out of use,

The first concept of contradiction can be seen in the work of Heraclitus of Ephesus (~500 BCE). Heraclitus is often credited as one of the first philosophers to see contradiction not as a logical flaw but as the very engine of change in nature and society (Wheelwright, 1968). His famous fragments proclaim that 'strife is the father of all things' (DK22B53) and 'you cannot step twice into the same river' (DK22B12), stressing that everything exists in a state of flux and that unity emerges through the tension of opposites. Heraclitus's state of flux represents a particular case of the unity of opposites, pointing to the way things are both the same and not the same over time. He depicts key opposites that are interconnected. For Heraclitus, the world is in an ongoing process of becoming, shaped by the constant interplay and conflict of contradictory forces.

Socrates and Plato (427–347 BCE) were other philosophers of Ancient Greece who spoke about the concept of contradiction. Plato's records of Socratic dialogues reframe contradiction within the dialectical method (Kahn, 1981). In dialectical method apparent inconsistencies or opposing claims are examined through structured questioning with the goal of arriving at clearer definitions, uncovering assumptions, and ultimately moving the interlocutors closer to truth or justified belief. In these dialogues, contradiction appears when ideas are tested against each other through questioning. While this highlights the value of critical examination, contradiction itself is treated as a problem to be resolved in the pursuit of truth rather than as a necessary principle of transformation. Contradictions are employed as a tool in philosophical inquiry and self-discovery. By highlighting contradictions, Socrates aimed to reveal deeper truths about reality and virtue, pushing his interlocutors to refine their understanding.

Plato's philosophy is grounded in the distinction between the ideal and the material realms. In the dialogues, sensible objects are treated as imperfect and mutable copies of immutable and eternal forms (or ideas), such as justice, beauty or equality. Knowledge of these forms is accessible not through empirical observation but through dialectical reasoning, which enables the mind to ascend from shifting appearances to the stable intelligible order. This framework introduces a powerful dualism between material manifestations and ideal essences, a distinction that proved influential across the history of philosophy. It also establishes a template for later debates about the relationship between the world as it is encountered and the world as it ought to be understood, a tension that resurfaces in different guises in modern dialectical materialism.

Another philosopher from Ancient Greece, Aristotle (384–322 BCE), took a decisive step in formalising contradiction as a central principle of logic. His law of non-contradiction in *Metaphysics Gamma* states that a thing cannot both be and not be at the same time and in the same respect (Seddon, 1988). According to Aristotle, contradiction operates on three levels: ontological (being), doxastic (belief) and semantic (meaning). For example, ontologically, a thing cannot have contradictory

especially *Second World*, they are relevant here for understanding the historical context in which cultural-historical theory circulated.

properties; doxastically, a person cannot believe contradictory things simultaneously; and semantically, a statement and its negation cannot both be true. Aristotle's influence shaped the idea that consistency is the foundation of rational thought (Massie, 2022). This principle became a cornerstone of Western logical reasoning: contradiction is to be avoided if an argument is to hold true.

Aristotle further developed the treatment of contradiction and change by embedding them within a broader account of causation, teleology and motion. Central to this system is the notion of the "unmoved mover", a first cause that initiates motion without itself being moved. While the unmoved mover is not a dialectical concept in the Hegelian sense, it introduces an important distinction between internal principles of change and external forces acting upon a system. For Aristotle, things possess intrinsic potentials that unfold toward their proper ends, yet these processes are conditioned by a hierarchy of causes that ultimately trace back to an external initiator. This framing is relevant for later discussions of contradiction because it foregrounds the question of whether change arises from internal tensions within a system or from forces acting upon it from without a distinction that remains salient in modern socio-historical and activity-theoretical analyses.

After centuries in which the logical view dominated, contradiction returned to the centre of philosophical thinking with the contributions of the German philosopher G.W.F. Hegel (1770–1831). Hegel did not intend to violate the law of non-contradiction but fundamentally reinterpreted it: for him, contradiction was not an error but the driving force behind the movement of concepts and reality itself. Hegel's approach, then, to overcoming the age-old Kantian contradiction between Sensation (intuition) and Reason (concept) holds that these opposites are in fact in unity and operate together as an outcome of human history (Blunden, 2021). Hegel argued that every idea contains within itself its own negation, leading to conflict and ultimately to an overcoming of this; this dialectical movement is now described using the terms 'thesis', 'antithesis' and 'synthesis'. Seen in this light, contradiction became a positive, generative dynamic that explains how development unfolds. In Hegel, dialectics is primarily conceived as an internal movement of thought, whereby concepts unfold through negation and sublation within the mind of a reflective individual. Although this process is historically and socially situated in Hegel's broader system, the dialectical progression itself can be read as something that an individual thinker can, in principle, engage with alone. Feuerbach challenged this orientation by shifting dialectics out of the solitary mind and into the sphere of human relations. As he insisted, "The true dialectic is not a monologue of the solitary thinker with himself. It is a dialogue between 'I' and 'You'" (Feuerbach, 2013). Here, contradiction is not simply a conceptual tension within thought, but a lived tension between persons, where understanding emerges through encounter, recognition and dialogue. Introducing Feuerbach at this point foregrounds the intersubjective dimension of contradiction and prepares the ground for later materialist and socio-historical accounts.

The German political theorists Karl Marx (1818–1883) and Friedrich Engels (1820–1895) adapted Hegel's dialectical method to the analysis of material social reality (Marx & Engels, 1967). Marx 'turned Hegel on his head' (Korsch, 2013), arguing that contradictions are not merely mental or conceptual but grounded in material conditions, such as the clash between the productive forces of industry and the relations of production in capitalist economies. For Marx, the alienation of labour from its product is not an abstract notion but a very real contradiction that labourers encounter in their lived experience.

Marx and Engels saw contradictions as inherent to capitalism, driving historical change and leading to its eventual overthrow. They identified key contradictions such as the conflict between the bourgeoisie (owners of capital) and the proletariat (workers), and the tendency of capitalist production to overproduce, leading to crises. According to their theory of historical materialism, these contradictions will ultimately lead to a revolutionary transformation of society.

Marx considered that it was possible to differentiate between principal and subordinate contradictions within a system, and that, in any activity system, at every stage in the development of a process, there is only one principal contradiction that plays the leading role. According to Marx, once this contradiction is identified it is possible to solve all problems (Althusser, 1969).

A central distinction in Marxist dialectical materialism elaborated by Marx and Engels, and later systematised by Lenin is that between internal and external contradictions. The character of a contradiction depends on the specific nature of the opposing sides and on the conditions in which their interaction takes place. Internal contradictions arise from the interaction of opposite sides within a given system, for example, within a species, within an organism or within a society. External contradictions involve the interaction of opposites belonging to different systems, for instance, between society and nature, or between an organism and its environment.

The universality of contradiction has a twofold meaning: firstly, contradiction exists in the process of development of all things; secondly, within the development of each thing, a movement of opposites exists from beginning to end. Ultimately, the decisive contradictions in development are the internal ones. There is internal contradiction in every single thing, hence its motion and development. Contradictoriness within a thing is the primary cause of its development, while its interrelations and interactions with other things are secondary causes.

Changes in society are chiefly due to the development of its internal contradictions. For example, the contradiction between the productive forces and the relations of production, the contradiction between classes, and the contradiction between the old and the new. The development of these contradictions pushes society forward and provides the impetus for the replacement of the old society by the new.

The sciences themselves are differentiated according to the particular contradictions inherent in their respective objects of study. The contradiction peculiar to a certain field of phenomena constitutes the object of study for a specific branch of science: positive and negative numbers in arithmetic; action and reaction in mechanics (Newton laws); positive and negative electricity in physics; dissociation and combination in chemistry; forces of production and relations of production, as well as classes and class struggle, in social science; offence and defence in military science; idealism and materialism, as well as the metaphysical outlook and the dialectical outlook, in philosophy. All these are the focus of different branches of science because each branch has its own particular contradiction and essence.

Without grasping the universality of contradiction, we cannot discover the universal cause or basis for the movement or development of things; but without studying the particularity of contradiction, we cannot determine the specific essence of a thing that differentiates it from others, nor discover the particular cause or basis for its movement or development, meaning that we would be unable to distinguish one thing from another or define the fields of science.

These contradictions cannot be treated in the same way since each has its own particularity. Moreover, the two aspects of each contradiction cannot be treated identically since each aspect has its own characteristics. It is crucial to understand what specific position each aspect occupies, what concrete forms it assumes in its interdependence and struggle with its opposite, and what concrete methods are employed in their interaction—both when the two aspects are interdependent and contradictory, and after their interdependence breaks down. Failing to understand both aspects of a contradiction means looking at a problem one-sidedly, seeing either the part or the whole, either the trees or the forest (Lenin, 1959).

The fundamental contradiction in the process of development of a thing, and the essence of the process determined by this fundamental contradiction, will not disappear until the process is complete. However, in a lengthy process, the conditions usually differ at each stage. Although the nature of the fundamental contradiction and the essence of the process remain unchanged, until a climax/revolution. In addition, among the numerous major and minor contradictions determined or influenced by the fundamental contradiction, some become intensified, some are temporarily or partially resolved or mitigated, and some new ones emerge. Therefore, the process is marked by stages. If people do not pay attention to the stages in the process of development of a thing, they cannot deal with its contradictions properly (Marx, 1859/1978).

The Russian politician Vladimir Lenin called for the concrete analysis of concrete conditions (1959). This means that in studying the particularity of any kind of contradiction, we must not be subjective nor arbitrary but must analyse its material nature. Without concrete analysis, there can be no knowledge of the particularity of any contradiction.

This materialist reinterpretation of contradiction shaped how later thinkers applied dialectics in the social sciences, especially in the Soviet context. Lev Vygotsky (1896–1934) drew on Marxist dialectics to analyse psychological development. Although he did not formalise the term 'contradiction' in the same way as Marx or Hegel, his work is deeply dialectical: he showed that learning and development occur through conflicts between actual and potential capacities, which is the core of his famous ZPD (Cole, 1996, 1998, 2005, 2010; Cole & Engeström, 1993).

Vygotsky influenced his fellow Russian psychologist, Aleksei Leontiev (1903–1979), whose emphasis on mediated action shifts the focus of analysis from the signs produced by an action to the action itself, which is mediated by tools (physical and conceptual) (1981). This necessitates an understanding of the specific tools used in different settings and how they are employed.

For the Russian philosopher Evald Ilyenkov (1924–1979), Leontiev's shift of focus towards action requires a deeper examination of the cultural and historical context that has shaped and standardised social actions over time (Ilyenkov & Korovikov, 1954). Ilyenkov himself was the first to introduce the concept of contradictions into what Leontiev called 'activity theory', and he viewed these as dynamic tensions that historically accumulate between opposing forces within an activity system (1974).

In turn, the work of Vygotsky, Leontiev and Ilyenkov influenced Finnish scholar Yrjö Engeström (1948–). Although Vygotsky laid the methodological foundations and Leontiev elaborated the structure of activity, it was Ilyenkov who reasserted the centrality of dialectical contradiction within Soviet philosophy. In his writings on the ideal, thinking and logical form, Ilyenkov argued that philosophy is "the science of scientific thinking, its laws and forms," emphasising that cognition cannot be understood apart from the material and practical conditions that give rise to it. This position reinforced a view of contradiction as the motor of thought and development, rather than as a logical error or a sign of conceptual failure. Engeström drew on this intellectual lineage and made contradiction an explicit analytical concept within activity theory, developing a practical typology of contradictions to understand how tensions within and between nodes of an activity system generate change and innovation.

In their 2011 paper, Yrjö Engeström and Annalisa Sannino argued that a contradiction is first and foremost a foundational philosophical concept and must not be equated with paradox, tension, inconsistency, conflict, dilemma or double bind. They stressed that these phenomena should be theoretically defined on their own, related to the concept of contradiction, and understood as its manifestations. They wrote that, 'Contradictions cannot be observed directly; they can only be identified through their manifestations' (2011, p. 368). Therefore, 'contradictions are historically emergent and systemic phenomena' to which we have no direct access (2011, p. 371). Hence, we can only approach them through their manifestations, which can be viewed as articulations or constructs. In other words, contradictions become visible when practitioners articulate and construct them in words and actions (Karanasios et al., 2017).

Engeström (2011) further explains that contradictions are historically accumulating structural tensions within and between activity systems. An activity system constantly works through tensions and contradictions within and between its nodes, and these contradictions manifest themselves in disturbances and innovative solutions. Subjects can experience contradictions at four levels: primary, secondary, tertiary and quaternary, depending on where they occur in the activity system (Cole & Engeström, 1993; Roth, 2004).

Contradiction exists in and runs through all processes from beginning to end; motion, things, processes and thinking all contain contradictions. To deny contradiction is to deny everything. This is a universal truth for all times and contexts, which admits of no exception, which is what gives contradiction its general characteristic of absoluteness.[2]

Yet, this general characteristic is always contained within every individual characteristic; without individual characteristics, no general characteristic can exist. It is precisely because each contradiction is particular that individual characteristics arise. All individual characteristics exist conditionally and temporarily, and are therefore relative. Yet if all individual characteristics were removed, no general characteristic would remain (Lenin, 1959) (Table 2.1).

How the Concept of Contradiction in Cultures of the Non-Western Resonates with CHAT

While Western theories of contradiction often focus on structural or systemic tensions, philosophies from the Global South[3] tend to locate contradiction within the moral, spiritual and existential dimensions of human life. Concepts from these cultures exemplify how the synthesis of contradictions is deeply embedded within cultures around the world.[4]

In the Indian context, contradictions are understood as dialectically entailed opposites, expressed as P and –P: two opposing determinations that do not merely stand in logical negation, but rather exist *because of* one another. The presence of one entails the other, and their interdependence generates movement and qualitative

---

[2] The French philosopher Jacques Derrida (1930–2004) criticised the value of contradictions, proposing that binary opposites cannot hold and meaning must be deferred. While not focusing specifically on contradictions in the Hegelian or Marxian sense, Derrida's work is useful in highlighting how one can destabilise fixed ways of knowing. Whereas Derrida undermines binary structures to show the impossibility of stable synthesis, cultural-historical approaches treat contradictions as generative tensions that can be worked through in practice to produce new forms of activity. For this reason, deconstruction's epistemological critique does not translate into a methodology for studying systemic transformation in education, and thus remains outside the scope of the developmental, dialectical concerns addressed here.

[3] In this book, the term *Global South* is used to refer to regions in Africa, Latin America, parts of Asia and the Middle East that share historical experiences of colonialism, uneven development and structural marginalisation within global knowledge economies. The term *Global North* refers broadly to Europe, North America, Australia and parts of East Asia that are economically advantaged and dominant in the production and circulation of academic knowledge. These terms are preferred over older geopolitical labels (e.g. "Third World" or "developing countries") because they are less hierarchical and more widely accepted by scholars in the South themselves. When the term *Western* is used, it refers specifically to intellectual traditions rooted in European and North American thought, rather than to geographical or political blocs; in this sense, "Global North" and "Western" partially overlap but are not synonymous.

[4] It is also worth noting that scholars in the South sometimes use the acronym *WEIRD* (Western, Educated, Industrialised, Rich and Democratic) to describe Global North research contexts, highlighting that much psychological and educational research from these regions is based on unrepresentative populations. For these reasons, the terminology adopted here aims to respect how different communities prefer to describe themselves while maintaining analytical clarity.

Table 2.1 A brief history of contradictions in Western and Soviet philosophy

Philosopher	Period	Key idea of contradiction	Role of contradiction
Heraclitus of Ephesus	~500 BCE	The struggle of opposites is the source of all becoming.	Contradiction is natural and necessary; everything exists through tension and conflict.
Said by Socrates and written down by Plato	~427–347 BCE	Dialectics is a method of investigating truth through opposing arguments.	Contradiction appears in dialogue. It is used to test ideas but is not treated as a positive principle of change.
Aristotle	~384–322 BCE	According to his law of non-contradiction, a statement cannot be true and false at the same time.	Contradiction is a logical fault, something to avoid in rigorous argument.
Immanuel Kant	Eighteenth-century enlightenment philosophy	Contradiction appears as an issue within reason itself, most notably in the antinomies, where reason generates opposing but rationally defensible claims.	Contradiction reveals the limits of pure reason and motivates the critical project; it forces a distinction between the phenomenal and noumenal and regulates what reason can legitimately claim.
Hegel	1770–1831	Contradiction is an inner feature of concepts; it is the source of movement and development.	Contradiction is the core driver of dialectical logic; all concepts contain their own negation, which produces change.
Ludwig Feuerbach	Nineteenth-century post-Hegelian materialism	Contradiction arises from the alienation between human essence and its religious projections, reflecting a split between real, sensuous humanity and abstract theological constructs.	Contradiction drives critique and demystification: by resolving the split, Feuerbach relocates abstract religious predicates back into human, sensuous existence, grounding philosophy in material human life.
Marx and Engels	Nineteenth century	Contradiction is found in social and economic structures.	Contradiction drives historical change, e.g. the conflict between the forces of production and the relations of production.
Vygotsky	1896–1934	Dialectics explain learning.	Contradiction between actual and potential development creates a space for learning (ZPD).
Leontiev	1903–1979	Contradictions exist between individual action and collective activity.	Contradictions explain tensions between motives, goals, means and conditions of human activity.

(continued)

Table 2.1 (continued)

Philosopher	Period	Key idea of contradiction	Role of contradiction
Ilyenkov	Mid-twentieth century Soviet philosophy	Contradiction is an objective feature of reality and thought, not a logical mistake; it is inherent in the movement from the concrete to the abstract and back again.	Contradiction is the driving force of thinking and development, compelling the formation of richer concepts and linking practical activity with the ideal.
Engeström	1948–	A typology of contradictions can be perceived in activity systems: primary, secondary, tertiary and quaternary.	Contradictions are the engine of expansive learning and organisational transformation.

change. Contradiction therefore is not viewed as error or inconsistency, but as a constitutive feature of activity and development.

This interpretation resonates with the cultural-historical tradition, where contradictions are treated as sources of systemic tension and drivers of expansive transformation rather than as problems to be eliminated.

In Chinese philosophy, *yin-yang* represents the dynamic interplay of opposing yet complementary forces that together sustain balance and harmony in the universe. Rather than framing contradiction as a conflict to be eradicated, this worldview recognises tension as a generative condition necessary for transformation and renewal.

The Chinese politician Mao Tse-tung (1937/1969:333) was likely drawing on this when he wrote about the two aspects of a contradiction:

> Of the two contradictory aspects, one must be principal and the other secondary. The principal aspect is the one playing the leading role in the contradiction. The nature of a thing is determined mainly by the principal aspect of a contradiction, the aspect which has gained the dominant position. But this situation is not static; the principal and the non-principal aspects of a contradiction transform themselves into each other and the nature of the thing changes accordingly.

The Chinese philosopher Confucius (551–479 BCE) articulated an epistemic stance grounded in humility and self-awareness: "When you know a thing, to hold that you know it; and when you do not know a thing, to allow that you do not know it – this is knowledge." This formulation positions knowledge not merely as the accumulation of information, but as the capacity to recognise and admit the limits of one's understanding. Confucius's statement implicitly recognises a dialectical tension between knowing and not-knowing. Rather than treating this as an error, he frames it as a condition for genuine understanding. This resonates with the CHAT perspective, where contradictions understood as tensions between opposing determinations create developmental possibilities. Admitting not-knowing allows contradictions to surface and be worked on, rather than denied or prematurely resolved.

Similarly, in African philosophy, *ubuntu*, which means that a person becomes a person through other people, expresses a dialectic between individuality and collectivity. It affirms that a person's humanity is realised through relationships with others, thereby foregrounding interdependence as a formative dimension of human development.

This resonates with Marx's idea that consciousness arises only in relation to the consciousness of others (Lukacs, 1972), a view later elaborated by Vygotsky who was critical that traditional views of higher mental functions did not distinguish between 'the biological and the social' in mental development (1997, p. 2). Vygotsky's emphasis that development begins on the social plane, through interaction with others, aligns closely with *ubuntu*. Both perspectives position social relations as foundational to the emergence of the individual, highlighting how the child develops through the influence and participation of people within their environment.

This synthesis reconciles the tension between self and community, autonomy and interdependence, by privileging relationality and mutual becoming over separation. Both philosophical systems thus reveal an understanding of contradiction as a creative and developmental principle, one that aligns closely with the dialectical foundations of CHAT.

A similar current is evident in Middle Eastern perspectives, such as those expressed in the writings of Kahlil Gibran. In *The Prophet*, Gibran captures the coexistence of opposites through poetic paradoxes that mirror dialectical harmony. He writes, 'beauty is life when life unveils her holy face. But you are life and you are the veil' (1993, p. 83). In this he encapsulates the inseparability of the visible and the hidden, the finite and the infinite. Here, contradiction is not a site of division but of revelation: the veil that conceals also discloses meaning. This resonates with the *yin-yang* notion of unity within duality and the *ubuntu* understanding of self-realisation through relational interdependence. Gibran's vision thus extends the non-Western dialectical tradition by portraying contradiction as an essential feature of being, where beauty, truth and life emerge through the continual unveiling and veiling of opposites.

It is evident that the synthesis of contradictions is not confined to Western dialectical traditions but forms a core dimension of human meaning-making across diverse cultural epistemologies. Collectively, these traditions—Western (Russian/Soviet tradition), Chinese, African and Middle Eastern—affirm that synthesising contradictions is a universal mode of thought that nurtures harmony, transformation and the deepening of human understanding.

Contradictions in Dialectics and Dialogics

Dialectics, for Hegel, is a logical method of philosophical inquiry where contradictions are used to find their own solution (see Tables 2.2 and 2.3). In Hegel's discussion of the master and slave dialectic (2004), he argues that self-awareness develops in our relations with others. He writes that the power and control that a master has

Table 2.2 Summary of the difference between dialectics and contradictions

Contradictions	Dialectics
Concrete tensions, clashes or opposing forces	A method, logic and approach to understand change
Concerns what actually happens	Concerns how to think
Are the driving forces that make dialectics work	Looks at how contradictions interact

Table 2.3 Contradiction as conceptualised in three distinct periods

	Classical philosophy	Dialectical philosophy	Contemporary philosophy (CHAT)
Main idea	Contradiction as a logical fault or rhetorical tool	Contradiction as an inherent driver of development and transformation	Contradiction as an analytical lens for studying and transforming activity systems
Key figures	Heraclitus (struggle of opposites), Plato (ideal and the material forms), Aristotle (law of non-contradiction)	Hegel (logical contradiction in concepts), Marx (material contradiction in social structures)	Vygotsky (dialectical roots in psychology), Leontiev (activity theory), Engeström (expanded CHAT with levels of contradiction)
Role of contradiction	To be avoided or resolved for logical consistency	To be embraced as a necessary tension that drives change (thesis–antithesis–synthesis; class conflict)	To be identified, analysed, and used as a practical tool for understanding breakdowns and designing solutions in education and work
Nature of contradiction	Often externalised or treated as error	Internal, generative, and inherent to all development	Embedded in practice: shows up through tensions, breakdowns and double binds
Key contribution	Sets foundations for logical clarity and debate (non-contradiction principle)	Provides a method for explaining historical and social change	Provides a practical framework for empirical research, learning design and organisational change
Typical example	Aristotle: A cannot be B at the same time.	Marx: conflict between productive forces and relations of production	Engeström: tension between new tools and old rules in an activity system

over a slave is, in fact, detrimental to the master attaining full self-awareness—as it is the slave who has access to learning because he works within the world, and it is the master who actually depends more on the slave than vice versa. Because of his focus on his ideas, Hegel's form of dialectics is characterised as dialectical idealism. From a Global South perspective, where slavery is a concrete historical system linked to colonialism and racial capitalism, such a metaphorical treatment is ethically incomplete. Hegel offers no pathway toward abolition, liberation or structural transformation, which illustrates the limitations of dialectical idealism when confronted with real historical oppression.

In contrast, Marx's dialectical materialism focuses on the concrete meaning of phenomena and real material relations (Sannino & Engeström, 2018). A central basis for dialectical materialism is the understanding of the importance of history in the development of society. For Marx, the contradictions that emerge between classes would eventually lead to revolution, pushing society to change. While Marx envisaged communism as the end of class antagonism—the conclusion of the "pre-history of human society"—this does not imply a static society, but rather a new form of development no longer determined by class struggle.

CHAT's dialectical approach draws more from Marx's dialectical materialism than Hegels' dialectical idealism. Vygotsky notes that, 'all scientific thinking moves by dialectics' (1983, p. 37) and he refers to dialectics in his work, both pointing to dialectical logic and explaining with the help of its laws.

Dialectics should not be confused with dialogics, a concept that is often associated with the Russian philosopher, Mikhail Bakhtin (1895–1975). Bakhtin wrote, 'Truth is not born nor is it to be found inside the head of an individual person, it is born between people collectively searching for truth, in the process of their dialogic interaction' (1984, p. 110). While the contradictions in a dialectical argument arise in most part from dialogic interaction, such interaction does not necessarily involve dialectics.

Dialogical pedagogy is easily confused with dialectical pedagogy because both forms of pedagogy rely on language and both accept that knowledge is a collective intersubjective endeavour. However, while dialogical pedagogy views differences as constitutive of meaning, dialectical pedagogy aims to overcome differences (Wegerif, 2011).

In this book, while we adopt an explicitly dialectical approach to learning, we also recognise the important impact of dialogue in a dialectical framework. Dialogue supports dialectical moments; it enables dialectical transitions in the process of development, allowing for the qualitative reorganisation of the whole system of mental functions as a result of such transitions (Veresov, 2015).

The dialectical nature of education for development can be seen if we conceive of contradictions as creating a disequilibrium in a system, forcing a change. In dialectical pedagogy the goal is to develop the student from a space of not knowing something to a space of new knowledge. The teacher's job is to stimulate argument that enables the student to see what it is that they do not know so that they can then learn this. It is in this sense, then, that dialectics is about building knowledge.

Dialectics is a broad method of thinking and understanding change. It is the logic of movement, development and transformation through opposing forces and it is a way of seeing the world as dynamic and contradictory (Leontiev, 1977, 1978, 1992). The kinds of questions one asks in a dialectical frame are:

- How does something become what it is?
- What are the opposing forces or tensions driving its development?
- How do these opposites interact, clash and become resolved as something new?

Contradictions cannot be seen except through their manifestations, such as when we note conflict in an argument between two speakers. Contradictions surface in a system through discursive means. Tensions, dilemmas and conflict are manifestations of contradictions; it is these we 'see' and track. Contradictions are the clashes within and between activity systems, also reflected in the conceptualizations of these systems by those who inhabit them as shown in Table 2.2.

Resolving contradictions is not a process of subsuming one position into its contradictory other. Rather it involves inventing a third new way of being and doing that is qualitatively different from the historically prior form. Yet change in any system—that is the overcoming of contradictions—can be progressive, leading to a new way of being and doing, or regressive, where the system reverts to a historically former version of itself that ignores the contradiction (as shown in Chaps. 6 and 7, and discussed in Chap. 8).

Nevertheless, the ancient philosophical problem of contradiction has developed into a modern methodological tool, supporting robust analysis and practical intervention in complex human systems (see Table 2.3).

The Role of Contradictions in Psychological Development

The CHAT concepts of *perezhivanie* and the dramatic frame allow deeper exploration of the role that contradictions play in psychological development. Although these concepts arise from Soviet cultural-historical theory, one can note in passing that aspects of *perezhivanie* resonate with earlier psychoanalytic attempts to theorise how individuals mediate external demands and internal impulses—for example, Freud's account of the superego as an internalised social voice and Jung's notion of the persona as a socially adaptive mask (Rubtsova, 2021; Rubtsova & Daniels, 2016). Similarly, the idea of a *dramatic frame* has antecedents in Erikson's account of psychosocial development, in which each stage is structured around a core conflict or collision of opposites (e.g., trust vs. mistrust, identity vs. role confusion). The CHAT perspective differs in its emphasis on socio-material activity and collective motives, but these brief comparisons help situate *perezhivanie* and drama within a broader intellectual genealogy concerned with how tension, conflict and contradiction shape the developing person.

The Experiential Prisms of Perezhivanie

Literally, in Russian, '*perezhivanie*' means 'experiencing', but in the context of developmental psychology it is a more complex concept. Vygotsky explain *perezhivanie* is the mediating link between the inner world of the individual and the surrounding reality, and therefore this central construct is closely connected to the social situation of development and the ZPD. *Perezhivanie* is an autonomous process through which the person relates to the world and attempts to resolve real-life problems. Vygostsky explains:

> A perezhivanie is a unit where, on the one hand, in an indivisible state, the environment is represented ... and on the other hand, what is represented is how I, myself, am experiencing this... So, in a perezhivanie we are always dealing with an indivisible unity of personal characteristics and situational characteristics, which are represented in the perezhivanie. (1994, p.342)

According to Smagorinsky (2011a, 2011b, p. 339), despite its theoretical importance, *perezhivanie* 'remains more a tantalizing notion than a concept with clear meaning', yet Vygotsky consistently worked to clarify and expand its developmental significance.

Already in his early writings (Vygotsky, 1986), had rejected the simplistic distinction between the individual and the social, observing that 'The social also exists where there is only one person with his individual *perezhivanie*' (p. 314). Later, he refined this argument by establishing a direct relationship between *perezhivanie* and development, proposing that the environment shapes development only through the child's *perezhivanie* of it (1998, p. 294). According to Vygotsky, *perezhivanie* provides a means to analyse the influence of the sociocultural environment not on the individual as such, but on the individual's process of development (1994, p. 343).

This perspective extends the developmental framework by rejecting the dualistic principle of reflection, which assumes a direct relationship between external stimuli and internal responses, and replacing it with the dialectical principle of refraction. No social factor has developmental significance in itself; it acquires significance only as it is refracted through the child's *perezhivanie* (Vygotsky, 1994, pp. 339–340). In this sense, *perezhivanie* functions as a psychological prism that determines how the environment affects development (Vygotsky, 1994, p. 341). The child is inseparable from the social situation, and the relationship between the individual and the environment is mediated by the individual's *perezhivanie* (Vygotsky, 1998, p. 294). The principle of refraction therefore demonstrates the dialectical relationship between the social and the individual: the social becomes the individual, but only those elements of the social world refracted through *perezhivanie* acquire developmental power.

Perezhivanie can also be understood as a dialectical contradiction in itself. It is a form of activity rather than being a single psychological function equivalent to memory, perception, thinking, imagination or emotion. Rather, these psychological functions, together with external object-oriented actions, participate in the realisation of *perezhivanie* in much the same way that they participate in the performance of any human activity. Their meaning and significance become clear only when viewed through the integrative work that *perezhivanie* performs: transforming psychological reality when external, practical action is no longer adequate to the situation. This aligns with Vygotsky's account of developmental crises, in which external circumstances and internal resources come into contradiction, giving rise to qualitative reorganisation rather than simple incremental change. In such moments, *perezhivanie* functions as both prism and processor: it refracts the situation into its component meanings, tensions and demands, but also reverses this analytic process by recombining these differentiated elements into a new psychological whole. The metaphor of the two prisms (see Fig. 2.1) captures this double movement, where

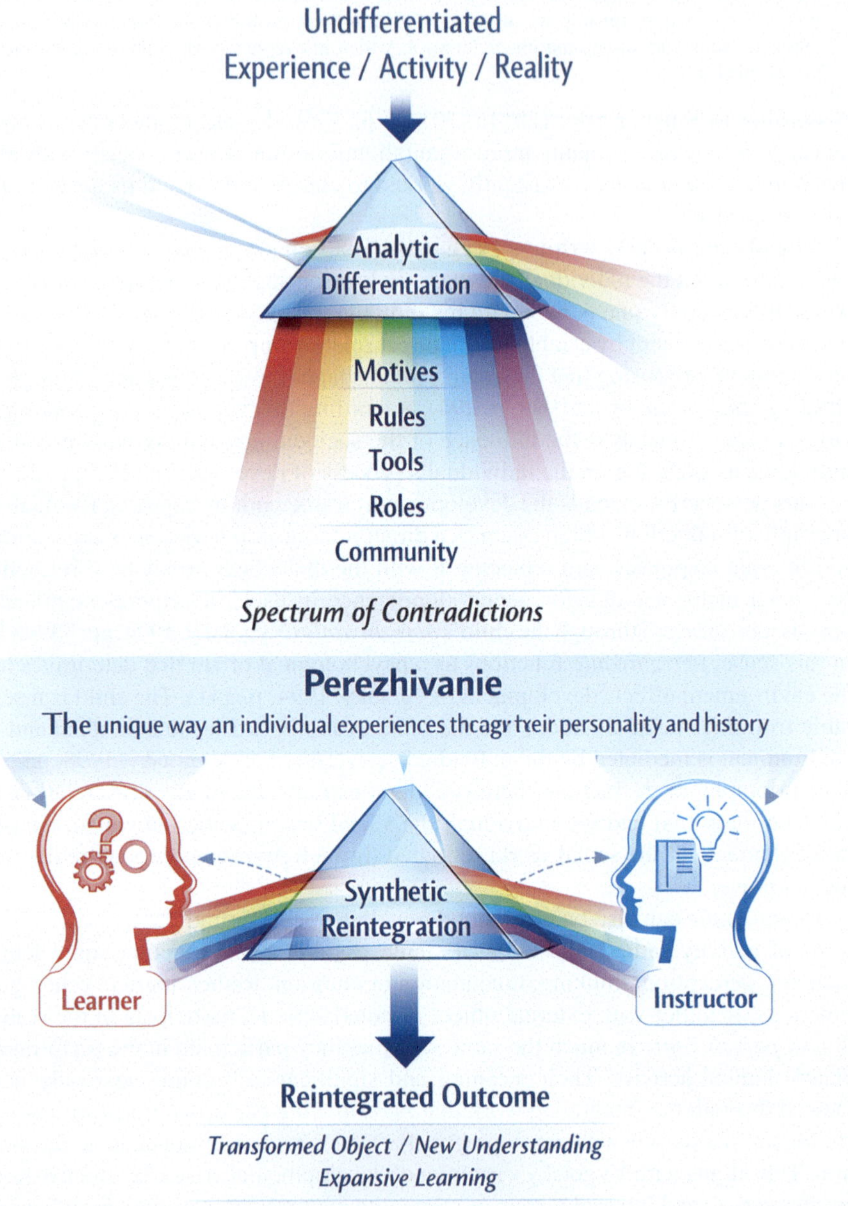

Fig. 2.1 *Perezhivanie* diagram

lived contradictions are made visible and then reintegrated, allowing development to proceed through qualitative transformation rather than mechanical adaptation.

In Fig. 2.1 at the very top you can find "Undifferentiated Experience / Activity / Reality". This represents the flow of everyday life, learning and teaching before it has been analysed or thematised. For a student, this might be the whole situation of studying Physics with an e-textbook in a particular institution, at a particular time, with particular peers and lecturers (see empirical study in Chap. 4). Everything is there at once: tools, rules, expectations, emotions, histories. It is "white light" in the sense that all the colours are present but not yet distinguishable.

The first prism is labelled "Analytic Differentiation". When experience passes through this prism it is refracted, and the white light of undifferentiated reality splits into a spectrum of colours. Psychologically and methodologically, this corresponds to the analytic moment in CHAT and qualitative research. The researcher (or reflective practitioner) begins to differentiate elements of the activity system: motives, rules, tools, roles, community, division of labour, objects, outcomes. What was previously fused in lived experience is pulled apart so that we can see where tensions lie. This is also where contradictions become visible: for example, between the promise of e-textbooks and students' lack of devices; between institutional rules and local practices; between lecturers' assumptions and students' capabilities. The coloured spectrum in the centre of the diagram is therefore a "spectrum of contradictions" in the sense that differentiation makes structural tensions and misalignments observable.

In the middle band the term "*Perezhivanie*" is placed as a bridge between the spectrum and the second prism. Vygotsky's concept of *perezhivanie* refers to the unique way in which an individual lives through, interprets and is affected by a situation. The environment does not act on the person directly; it acts through how the person experiences it, given their history, motives, identity and developmental level. In the diagram, *perezhivanie* sits between the two prisms to indicate that what passes into the second, synthetic prism is not the situation "as such" but the situation as refracted through the learner's and the lecturer's *perezhivanie*. Two head icons (learner and instructor) emphasise that there are at least two subjectivities involved. The same spectrum of contradictions can be experienced as opportunity, threat, frustration, curiosity or injustice, depending on each person's *perezhivanie*. For instance, the shift to e-textbooks (see empirical study in Chap. 4) might be experienced by a well-resourced, confident student as liberating, but by a student with no laptop and limited data as anxiety-provoking and exclusionary. The contradictions are structurally the same, but the lived experience of them differs.

The second prism is labelled "Synthetic Reintegration". Here the separated colours are recombined into a new form of "white light". This represents the integrative, developmental moment of the process. After contradictions have been made visible and experienced through *perezhivanie,* they can be worked on and potentially transformed. In CHAT terms, this is where expansive learning or object transformation can occur. For the learner, the spectrum of differentiated elements (difficulties, supports, rules, expectations, tools) is re-organised into a new understanding of the task, the subject, or their own learning. For the lecturer, their experience of institutional constraints, students' struggles, and technological affordances can be integrated into a revised pedagogy or advocacy strategy. The recombined "white light" at the bottom is therefore not a return to the original undifferentiated state; it is a qualitatively different unity—a reintegrated outcome such as a transformed object, a new shared understanding, or a reconfigured activity system.

The label "Reintegrated Outcome: Transformed Object / New Understanding / Expansive Learning" captures this. The outcome might be a redesigned way of using e-textbooks, a new division of labour between lecturer, student and tool, or a different way the student positions themselves as a learner. Importantly, the diagram implies a full dialectical cycle: undifferentiated experience; analytic differentiation (first prism); lived experience of contradictions (*perezhivanie*); synthetic reintegration (second prism); and a transformed whole that can itself become the starting point for a new cycle.

The Russian psychologist Fyodor Vasilyuk (1953–2017) has also written on *perezhivanie*, conceptualising it as a special form of activity that emerges in situations where a person can no longer achieve their main goals and life plans. For Vasilyuk, *perezhivanie* appears when ideals collapse (meaning that conceptual structures break down) and psychological reality requires transformation. Vasilyuk emphasises the sense of surviving when confronted with psychological crisis: the person passes through difficult experiences and overcomes intense feelings, and in doing so the person endures (Vasilyuk, 1984/1988).

As mentioned above Vygotsky introduced *perezhivanie* as a methodological unit that expresses the relation between a person and their environment. For Vygotsky, a situation affects the child not directly but through their *perezhivanie* of it. His focus was therefore relational and developmental: *perezhivanie* explained why the same environmental conditions could produce different developmental outcomes for different children. In this sense, *perezhivanie* was used to connect social situation, individual characteristics, and psychological development, emphasising how external conditions are refracted by the person's motives, affect, and age-related developmental structure.

By contrast, Fyodor Vasilyuk substantially reworked the concept. Writing in the 1970s–1980s, Vasilyuk understood *perezhivanie* not merely as a lens on environmental influence, but as a distinct form of psychological activity that comes to the foreground in critical life situations. For Vasilyuk, *perezhivanie* is what the person **does** when existing forms of action fail and the situation cannot be resolved externally. In these moments of crisis, *perezhivanie* becomes an internal, meaning-creating struggle through which the person regains psychological equilibrium and reorganises their relation to self and world. In short, where Vygotsky viewed *perezhivanie* as a unit of experience, Vasilyuk framed it as a process of experiential work.

It is important to mention while Vasilyuk builds on Vygotsky, he effectively shifts the locus of development inward, giving *perezhivanie* an existential, crisis-oriented and activity-based character that is largely absent from Vygotsky's original formulation. This makes Vasilyuk's account especially relevant for understanding how individuals confront suffering, loss, anxiety and other experiences in which development proceeds through inner work rather than external mastery.

The Dramatic Frame of Personality

Vygotsky's writings indicate that he conceived of the fundamental principle of higher mental functions, and therefore personality, as inherently social. Vygotsky viewed the interpsychological formation of higher mental functions not as the result of a neutral interaction between people, but as the result of a dramatic encounter: an emotionally charged conflictual engagement that shapes the personality participating in it (1989, p. 69).

The 'dramatic frame' of personality, the unique organisation and hierarchy of higher mental functions, emerges from the specific dramatic collisions a person experiences in life, and from how these collisions are resolved. Each individual's intrapsychological structure is the result of these lived dramas, internalised as part of their developmental trajectory.

For this reason, Vygotsky insisted that psychology must be conceptualised in terms of drama, rather than in terms of isolated processes (1989, p. 71). Overcoming the 'dramas of life' is what enables the creation of a unique personality. The personality, therefore, is not simply a collection of mental functions but a dynamic constellation shaped by the dramas it endures. Development is propelled by these dramatic contradictions: the tensions, conflicts and affective upheavals that compel the individual to transform.

This idea is directly rooted in Vygotsky's general genetic law of cultural development, which states that 'Every function in the cultural development of the child appears twice … first between people as an intermental category, then within the child as an intramental category' (1981, p. 163).

His repeated use of the term 'category' is significant. In pre-revolutionary Russian theatre, 'category' referred to a dramatic event, a collision of characters on the stage that generates the drama as a whole. Vygotsky, thoroughly familiar with theatrical language, uses this term deliberately to emphasise that the social relations from which higher mental functions originate are not ordinary but dramatic.

Vygotsky himself clarified the methodological importance of drama. In direct continuation of the general law, he states, 'One of the central principles of our work is the experimental unfolding of higher mental processes into the drama that happened between the people' (1983, p. 145). This shows that, in his theory, every higher mental function originates not simply in social interaction but in social drama—a contradiction enacted between people—and only later appears internally.

Understanding development as drama aligns with Vygotsky's concept of the social situation of development. A social situation becomes developmental when the environment presents new demands and challenges that the child is not yet capable of meeting. The child's current level of development collides with the demands of the social environment, and this contradiction is the true moving force of development. It exists neither in the child nor in the environment alone, but in the dramatic relationship between them.

This explains why development cannot be understood without acknowledging drama: the emotionally lived contradiction defines the starting point of the developmental process and the pathway along which the social becomes the individual.

Contradictions Within an Activity System

In CHAT, contradictions are seen as fundamental elements (Sannino, 2010). They play a vital role in driving the development of the system (Yamagata-Lynch, 2003, 2010). Contradictions are tensions within and between activity systems (Engeström, 2009) and do not necessarily refer to problems and conflicts. The theoretical significance of contradictions lies in their dual nature: they simultaneously represent sources of disturbance and instability while also serving as catalysts for creativity and expansive learning. They are considered inherent features of all activity systems rather than external disruptions.

According to Sannino and Engeström (2018), the various terms used by scholars to describe contradictions (such as 'paradox', 'tensions', 'inconsistencies', 'conflict', 'disturbances' and 'dilemma') to refer to phenomena that can be understood as manifestations of the same underlying concept, namely contradiction, even though these terms have their own specific meanings and should not be treated as direct synonyms.

The Nodes of an Activity System

As illustrated in Chap. 1, an activity system is usually represented in the form of an activity triangle. In the activity triangles developed and discussed in this chapter, the subject encompasses one or more people involved in the activity. The mediating/al artefact (means) encompasses the resource or resources that the subject draws on during the activity. The object is the motive of the activity (Stetsenko, 2005; 2021).

In the activity triangle depicted in Fig. 1.3, social practice is concerned with the nodes at the bottom of the triangle: rules, community and division of labour. The rules are the formal policies or informal norms of the organisation where activity is taking place, and which affect how the activity takes place. The community is the social group that the subject belongs to while engaging in the activity. The division of labour refers to how tasks are shared among these members of the community, which occurs in a space where there are asymmetrical power relations.

Mediated action is located in the activity system's central triangle and refers to mediated action between the subject, mediating artefact and object. By considering the role of a mediating artefact (means or tool) in mediating actions, researchers can gain insights into the ways in which social actions have become patterned and habitual within specific cultural and historical contexts.

Activity System Analysis

Activity system analysis examines the interrelationships of contradictions in order to reveal opportunities for system-wide improvement. This form of analysis is by nature holistic, and encourages a researcher to consider each node and the relationship between all the nodes when identifying contradictions and the way that they relate to each other.

However, a researcher should perhaps pay particular attention to the object of the activity system. The tensions that arise in a system affect subjects' ability to attain the object (the goal of the activity) by either taking a role as an obstacle, making it difficult for the subjects to attain the object, or by taking a role as an enabling influence, helping the subject to attain the object (Yamagata-Lynch, 2003). The subjects' interpretations of the object should be seen as dialogical, involving interaction with one another and with the meanings that have become associated with the activity over time. As Engeström and Toiviainen notes, contradictions transform an object into a 'moving, motivated, and future-generating target' (2010, p. 115).

A useful way to analyse contradictions and their consequences within an activity system is to examine how the tensions that emerge affect the subject's ability to attain the object, understood as the goal or motive of the activity. Rather than treating tensions as uniformly negative, this analytical approach attends to how they operate within the system: they may function as obstacles that impede progress towards the object, or they may operate as enabling influences that support and even accelerate movement towards the object. In other words, tensions do not merely signal disruption; they also reveal emerging possibilities for change as the system adjusts to new conditions or reorganises itself around new demands (Yamagata-Lynch, 2003). This way of thinking prepares the ground for a more explicit analysis of how particular contradictions manifest within educational activity systems, especially when learners and teachers negotiate new tools, rules and roles in pursuit of shared learning goals.

A Typology of Contradictions

As discussed above, Marx distinguished between principal and subordinate contradictions within a system, and Engels considered inner contradictions to be a greater influence than external contradictions. Engeström and Sannino (2011), Sannino and Engeström (2018) took this hierarchical approach to contradictions further by classifying them into four types.

According to Engeström and Sannino (2011), Sannino and Engeström (2018), contradictions can be primary, secondary, tertiary or quaternary based on their location within an activity system (Cole & Engeström, 1993; Roth, 2004). The primary, secondary, tertiary and quaternary contradictions occur within nodes of the activity triangle, between nodes of the activity triangle, between different versions of the activity triangle, and between neighbouring activity triangles, respectively (Engeström & Sannino, 2011; 2017; 2020; Sannino and Engeström, 2018) (see Fig. 2.2).

A primary contradiction may exist inside specific one node of CHAT, such as rules. A primary contradiction, pivotal in CHAT, keeps the activity system in a state

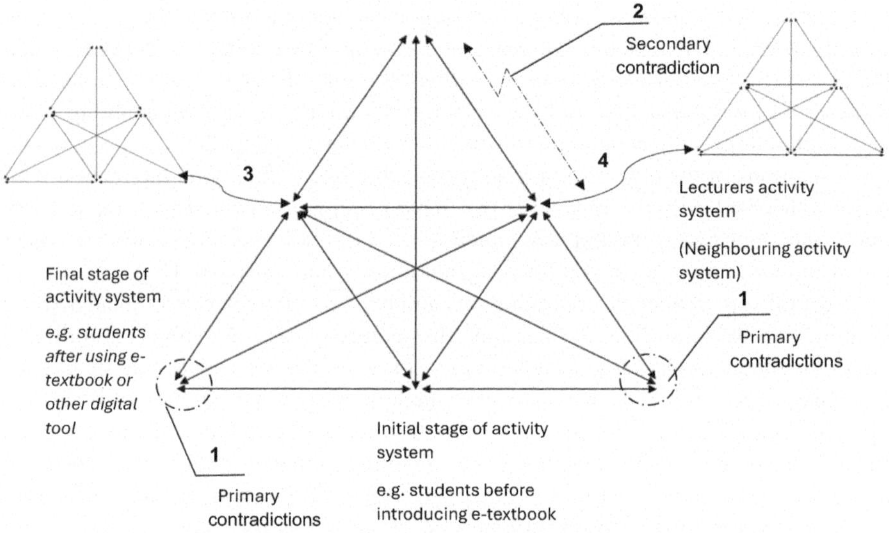

Symbol	*Academic & social* Description
Primary contradiction	This contradiction appear within elements
Secondary contradiction	This contradiction appear between elements of the system
3 Tertiary contradiction	This contradiction is between different versions of the same activity system. e.g. before and after or old and new model of teaching
4 Quaternary contradiction	This contradiction is between neighbouring activity system e.g. students – parents or students first year – students second year

Fig. 2.2 Engeström's primary, secondary, tertiary or quaternary contradictions in intersecting activity systems shown at two different times

of constant flux. It emerges in everyday contexts, taking various forms and influencing other levels of contradiction (Foot, 2014). The persistence of this primary contradiction is crucial; it not only remains a continual presence but also forms the basis for subsequent levels of contradiction.

Primary contradictions have the tendency to give rise to secondary contradictions that are specific to the particular conditions of the activity (Sannino, 2008a, 2008b, 2010). Even if resolutions are found for secondary contradictions, the underlying primary contradiction continues to exert its influence, perpetuating the dynamic nature of the activity system (Engeström, 2001).

While primary contradictions occur in the node (also referred to as the 'element') (Roth & Lee, 2007), a secondary contradiction arises between nodes. For example, integration of a new technology often clashes with established rules. Consequently, these contradictions lead to disturbances, conflicts and also novel attempts to modify the activity, turning it into an imperceptible battleground.

As an example of a secondary contradiction, consider the interaction between the mediating artefact, such as an e-textbook, and the corresponding academic community in South Africa. The e-textbook, being a written and static medium, contradicts the community's value of oral and dynamic knowledge transfer.

At the tertiary level, contradictions arise between an activity system and its potential forms in more advanced stages, or when the object of the older activity system clashes with the object of what is considered a more advanced activity system. For example, this can occur during the transition from traditional writing practices to the use of generative AI tools such as ChatGPT. Additionally, Roth and Lee (2007) suggest that tertiary contradictions can also occur between the object (motive) of the dominant activity and the object of a culturally advanced version of that activity.

Lastly, at the quaternary level, contradictions emerge when nodes of the central activity conflict with nodes of any neighbouring activity systems (Engeström, 2005, 2008, 2009; 2018; Roth & Lee, 2004, 2007), such as the interaction between students and lecturers.

Contradictions as Developmental Mechanisms

Contradictions and their resolution account for the motion of forces released in crystals, plants, animals, human beings, society and the whole universe. There are many ways of resolving contradictions and they depend on various conditions, including, in society, the character of the contesting parties. Sometimes one side of the contradiction perishes and the other triumphs. On other occasions, both sides perish, exhausting themselves in the struggle. There may also be a more or less prolonged compromise between the contestants. The resolution of a contradiction may be complete or partial, instantaneous or by stages, and it may result in either progress or regression (Horn, 2006). Yet, unless they are worked through, contradictions will not lead to progress; they are a necessary but not sufficient condition for development. Importantly, any resolution is provisional rather than final: once a contradiction is partially resolved, the resulting reorganisation of the activity system generates new conditions, and with them new tensions and contradictions. In this sense, contradictions may be transformed, displaced or revived in altered forms rather than eliminated altogether. Development therefore does not culminate in stasis, but proceeds through successive cycles of contradiction, resolution and renewed contradiction.

From a CHAT perspective, when contradictions intensify beyond the system's capacity for routine adjustment, they create conditions for expansive learning, a process through which activity systems reconceptualise their objects and reorganise their practices (as discussed in Chap. 1) (Engestrom, 2009). Thus rather than viewing contradictions as impediments to smooth functioning, CHAT conceptualises them as essential drivers of qualitative improvement. This developmental perspective challenges linear models of change that seek to eliminate problems through technical solutions. Instead, CHAT suggests that sustainable development requires engaging with contradictions dialectically. To engage with something dialectically means not treating tensions as errors to be removed, but as structural indicators of emerging possibilities. It entails analysing the opposing forces or motives within the activity system, understanding the historical and systemic conditions that produced them, and working through the tension in a way that generates a qualitatively new configuration rather than reverting to the previous order. In other words, the tension can be used productively to generate new forms of activity that transcend existing limitations while preserving valuable elements of current practice. This involves both a critical and a creative movement: identifying what is no longer adequate, and designing new objects, tools or roles that carry forward what is still useful.

In the context of education and pedagogy, engaging with contradictions dialectically means treating breakdowns in learning not as failures to be patched with technical fixes, but as opportunities to reconfigure the object and organisation of teaching. A useful illustration comes from mathematics education, where the introduction of calculators into classrooms produced a well-documented contradiction between procedural fluency and conceptual understanding. For many teachers, the calculator disrupted established routines that emphasised manual computation; for others, it created space to shift the object of activity from executing procedures to reasoning about quantitative relationships. If handled dialectically, the contradiction does not force a choice between "no calculators" and "full calculator dependence," but instead supports the emergence of new pedagogical practices that combine conceptual discussion, estimation strategies, and strategic use of tools. In such cases, the new form of activity both transcends earlier limitations (e.g., excessive time spent on mechanical paper-and-pencil algorithms) and preserves valuable elements of prior practice (e.g., number sense, estimation, and understanding of place value). This exemplifies how contradictions in educational settings, when engaged dialectically, can drive pedagogical innovation rather than mere adaptation.

An activity system is constantly navigating contradictions within and between its nodes, and the disruptions that surface are also opportunities for novel solutions (Engeström & Sannino, 2020). Contradictions contain the potential for the collective to search for new ways of doing and achieving 'what is not yet there' (Engeström, 2018; Engeström et al., 2010; Smagorinsky, 1995, 2011a, 2011b). As Engeström (1987) emphasises, tensions create opportunities and demand innovative solutions, and can ultimately lead to transformative shifts in the activities themselves (Engeström, 1987; Wertsch, 1990, 1991, 1998, 2003, 2007).

Change can be progressive or regressive, depending on whether it expands collective capabilities, deepens exclusion, or reinforces existing hierarchies. From a CHAT perspective, development is not assumed to be automatically progressive; rather, it emerges through the identification and working-through of contradictions in practice, which opens possibilities for expansive learning, new objects of activity, and more equitable forms of participation.

Conclusion

Owing to the progress of science and the increasing complexity of our systems of concepts—as well as the increasing diversity of people's social relationships—the philosophical and methodological importance of being able to identify and resolve contradictions is constantly growing. When we understand the principle of contradiction, this allows us to perceive the world as a place full of contradictions demanding to be known and resolved. It also encourages a flow of intellectual thought that prevents stagnation in any field, whether it be science, politics, art or education.

References

Althusser, L. (1969). Contradiction and overdetermination. In L. Althusser (Ed.), *For Marx* (pp. 87–128). New Left Books. (Original work published 1962).

Bakhtin, M. M. (1984). *Problems of Dostoevsky's poetics* (C. Emerson, Trans.). University of Minnesota Press.

Blunden, A. (2021). *Hegel, Marx and Vygotsky: Essays on social philosophy* (Vol. 195). Brill.

Cole, M. (1996). *Cultural psychology: A once and future discipline*. Harvard University Press.

Cole, M. (1998). Can cultural psychology help us think about diversity? *Mind, Culture, and Activity, 5*(4), 291–304. https://doi.org/10.1207/s15327884mca0504_4

Cole, M. (2005). Using cross-cultural psychology to design afterschool educational activities in different cultural settings. In W. Friedlmeier, P. Chakkarath, & B. Schwarz (Eds.), *Culture and human development* (pp. 53–71). Psychology Press.

Cole, M. (2010). Cultural–historical activity theory. In N. K. Denzin & Y. S. Lincoln (Eds.), *The sage handbook of qualitative research* (3rd ed., pp. 141–160). Sage.

Cole, M., & Engeström, Y. (1993). A cultural-historical approach to distributed cognition. In G. Salomon (Ed.), *Distributed cognitions* (pp. 1–46). Cambridge University Press.

Engeström, Y. (1987). *Learning by expanding: An activity theoretical approach to developmental research*. Orienta-Konsultit.

Engeström, Y. (2001). Expansive learning at work: Toward an activity theoretical reconceptualization. *Journal of Education and Work, 14*(1), 133–156. https://doi.org/10.1080/13639080020028747

Engeström, Y. (2005). *Developmental work research: Expanding activity theory in practice*. Lehmanns Media.

Engeström, Y. (2008). *From teams to knots: Activity-theoretical studies of collaboration and learning at work*. Cambridge University Press.

Engeström, Y. (2009). The future of activity theory: A rough draft. In A. Sannino, H. Daniels, & K. D. Gutiérrez (Eds.), *Learning and expanding with activity theory* (pp. 303–328). Cambridge University Press. https://doi.org/10.1017/CBO9780511809989.020

Engeström, Y. (2011). From design experiments to formative interventions. *Theory & Psychology, 21*(5), 598–628.

Engeström, Y. (2018). *Expertise in transition: Expansive learning in medical work*. Cambridge University Press.

Engestrom, Y. R. J. Ö. (2009). From learning environments and implementation to activity systems and expansive learning. *Actio: An International Journal of Human Activity Theory, 2*, 17–33.

Engeström, Y., & Toiviainen, H. (2010). Co-configurational design of learning instrumentalities: An activity theoretical perspective. In *Learning Across Sites* (pp. 45–64). Routledge.

Engeström, Y., Kajamaa, A., Kerosuo, H., & Laurila, P. (2010). Process enhancement versus community building: Transcending the dichotomy through expansive learning. In *Activity theory and fostering learning: Developmental interventions in education and work* (pp. 1–28). Center for Human Activity Theory, Kansai University.

Engeström, Y., & Sannino, A. (2011). Discursive manifestations of contradictions in organizational change efforts: A methodological framework. *Journal of Organizational Change Management, 24*(3), 368–387. https://doi.org/10.1108/09534811111132758

Engeström, Y., & Sannino, A. (2017). Studies of expansive learning: Foundations, findings and future challenges. In H. Daniels, M. Fleer, & P. Cole (Eds.), *The Cambridge handbook of cultural-historical psychology* (pp. 200–226). Cambridge University Press.

Engeström, Y., & Sannino, A. (2020). Toward a Vygotskian perspective on transformative agency for social change. In *Revisiting Vygotsky for social change* (pp. 87–109). Peter Lang.

Feuerbach, L. (2013). Principles of the philosophy of the future (Z. Hanfi, Trans.). In Z. Hanfi (Ed. & Trans.), *The fiery brook: Selected writings* (pp. 175–246). Verso Books. (Original work published 1843)

Foot, K. A. (2014). Cultural-historical activity theory: Exploring a theory to inform practice and research. *Journal of Human Behavior in the Social Environment, 24*(3), 329–347. https://doi.org/10.1080/10911359.2013.831011

Gibran, K. (1993). *The prophet*. Penguin Books.

Hegel, G. W. F. (2004). *Phenomenology of Spirit* (A. V. Miller, Trans.). Oxford University Press. (Original work published 1807).

Horn, L. R. (2006). Contradiction. In K. Brown (Ed.), *Encyclopedia of language and linguistics* (2nd ed., pp. 597–600). Elsevier.

Ilyenkov, E. V., & Korovikov, V. I. (1954). *Theses on the interconnection of philosophy and knowledge of nature and society in their historical development*. Moscow State University.

Kahn, C. H. (1981). Did Plato write Socratic dialogues? *The Classical Quarterly, 31*(2), 305–320. https://doi.org/10.1017/S0009838800003337

Karanasios, S., Riisla, K., & Simeonova, B. (2017). Exploring the use of contradictions in activity theory studies: An interdisciplinary review. Paper presented at the European Group for Organizational Studies Colloquium, Copenhagen, Denmark.

Korsch, K. (2013). *Marxism and philosophy*. Verso Books.

Lenin, V. I. (1959). *Philosophical notebooks* (A. P. Shestopal & J. L. Dickinson, Eds. & Trans.). International Publishers.

Leontiev, A. N. (1977). The dialectical method in the psychology of memory. *Soviet Psychology, 16*(1), 53–69. https://doi.org/10.2753/RPO1061-0405160153

Leontiev, A. N. (1978). *Activity, consciousness, and personality*. Prentice-Hall.

Leontiev, D. A. (1992). Joint activity, communication, and interaction. *Journal of Russian & East European Psychology, 30*(2), 43–58. https://doi.org/10.2753/RPO1061-0405300243

Lukács, G. (1972). *History and class consciousness: Studies in Marxist dialectics*. MIT Press.
Mao, Z. (1937/1969). On contradiction. In *Selected works of Mao Tse-tung* (Vol. 1, pp. 311–347). Foreign Languages Press. (Original work published 1937).
Marx, K. (1859/1978). Preface to *A contribution to the critique of political economy*. In R. C. Tucker (Ed.), *The Marx-Engels reader* (2nd ed., pp. 3–6). Norton.
Marx, K., & Engels, F. (1967). *The German ideology: Parts I & III* (R. Pascal, Ed.). International Publishers. (Original work published 1932).
Massie, P. (2022). Contradiction, being, and meaning in Aristotle's *Metaphysics* gamma. *Philosophies, 7*(2), 24. https://doi.org/10.3390/philosophies7020024
Roth, W.-M. (2004). Activity theory and education: An introduction. *Mind, Culture, and Activity, 11*(1), 1–8. https://doi.org/10.1207/s15327884mca1101_1
Roth, W.-M., & Lee, S. (2004). Science education as/for participation in the community. *Science Education, 88*(2), 263–291. https://doi.org/10.1002/sce.10113
Roth, W.-M., & Lee, Y.-J. (2007). Vygotsky's neglected legacy: Cultural-historical activity theory. *Review of Educational Research, 77*(2), 186–232. https://doi.org/10.3102/0034654306298273
Rubtsova, O., & Daniels, H. (2016). The concept of drama in Vygotsky's theory: Application in research. *Cultural-Historical Psychology, 12*(3).
Rubtsova, O. V. (2021). Experimenting with roles in adolescence: Applying drama for constructing the zone of proximal development. *Cultural-Historical Psychology, 17*(2).
Sannino, A. (2008a). Experiencing conversations: Bridging the gap between discourse and activity. *Journal for the Theory of Social Behaviour, 38*(3), 267–291. https://doi.org/10.1111/j.1468-5914.2008.00371.x
Sannino, A. (2008b). From talk to action: Experiencing interlocution in developmental interventions. *Mind, Culture, and Activity, 15*(3), 234–257. https://doi.org/10.1080/10749030802186769
Sannino, A. (2010). Teachers' talk of experiencing: Conflict, resistance and agency. *Teaching and Teacher Education, 26*(4), 838–844. https://doi.org/10.1016/j.tate.2009.10.021
Sannino, A., & Engeström, Y. (2018). Cultural-historical activity theory: Founding insights and new challenges. *Cultural-Historical Psychology, 14*(3), 43–56. https://doi.org/10.17759/chp.2018140305
Seddon, F. A. (1988). *The principle of contradiction in Aristotle's metaphysics gamma* (Doctoral dissertation). Duquesne University.
Smagorinsky, P. (1995). The social construction of data: Methodological problems of investigating learning in the zone of proximal development. *Review of Educational Research, 65*(3), 191–212. https://doi.org/10.2307/1170682
Smagorinsky, P. (2011a). *Vygotsky and literacy research: A methodological framework*. Sense Publishers.
Smagorinsky, P. (2011b). Vygotsky's stage theory: The psychology of art and the actor under the direction of Perezhivanie. *Mind, Culture, and Activity, 18*(4), 319–341. https://doi.org/10.1080/10749039.2010.518300
Stetsenko, A. (2005). Activity as object-related: Resolving the dichotomy of individual and collective planes of activity. *Mind, Culture, and Activity, 12*(1), 70–88. https://doi.org/10.1207/s15327884mca1201_6
Stetsenko, A. (2021). Scholarship in the context of a historic socioeconomic and political turmoil: Reassessing and taking stock of CHAT. Commentary on Y. Engeström and A. Sannino "from mediated actions to heterogenous coalitions: Four generations of activity-theoretical studies of work and learning". *Mind, Culture, and Activity, 28*(1), 32–43. https://doi.org/10.1080/10749039.2021.1874419
Vasilyuk, F. E. (1984/1988). *The psychology of experiencing: An analysis of how critical situations are dealt with* (R. Goodman, Trans.). Progress Publishers. (Original work published 1984).
Veresov, N. (2015). Perezhivanie in cultural–historical theory: The concept and its content. Keynote lecture presented at the Australia-Brazil ISCAR Symposium on Perezhivanie, Monash University.

Vygotsky, L. S. (1981). The development of higher forms of attention in childhood. In J. V. Wertsch (Ed.), *The concept of activity in Soviet psychology (pp. 191–240)*. Sharpe. (Original work published 1929).

Vygotsky, L. S. (1983). *The collected works of L. S. Vygotsky* (Vol. 1). Springer.

Vygotsky, L. S. (1986). *Thought and language*. MIT Press.

Vygotsky, L. S. (1989). Concrete human psychology. *Soviet Psychology, 27*(2), 53–77.

Vygotsky, L. S. (1994). The problem of environment. In R. W. Rieber & A. S. Carton (Eds.), *The collected works of L. S. Vygotsky* (Vol. 3, pp. 339–347). Springer.

Vygotsky, L. S. (1997). *Educational psychology* (R. W. Rieber, Ed.). Plenum Press.

Vygotsky, L. S. (1998). *The collected works of L. S. Vygotsky* (Vol. 5). Plenum Press.

Wegerif, R. (2011). From dialectic to dialogic. In B. E. Davis & B. Francis (Eds.), *Theories of learning and studies of instructional practice* (pp. 201–221). Springer.

Wertsch, J. V. (1990). Dialogue and dialogism in a socio-cultural approach to mind. In I. Marková & K. Foppa (Eds.), *The dynamics of dialogue* (pp. 62–82). Harvester Wheatsheaf.

Wertsch, J. V. (1991). A sociocultural approach to socially shared cognition. In L. B. Resnick, J. M. Levine, & S. D. Teasley (Eds.), *Perspectives on socially shared cognition* (pp. 85–100). American Psychological Association. https://doi.org/10.1037/10096-004

Wertsch, J. V. (1998). *Mind as action*. Oxford University Press.

Wertsch, J. V. (2003). Commentary on: Deliberation with computers: Exploring the distinctive contribution of new technologies to collaborative thinking and learning. *International Journal of Educational Research, 39*(8), 899–904. https://doi.org/10.1016/j.ijer.2004.11.012

Wertsch, J. V. (2007). National narratives and the conservative nature of collective memory. *Neohelicon, 34*(2), 23–33. https://doi.org/10.1007/s11059-007-2003-9

Wheelwright, P. (1968). *Heraclitus*. Colchis Books.

Yamagata-Lynch, L. C. (2003). Using activity theory as an analytic lens for examining technology professional development in schools. *Mind, Culture, and Activity, 10*(2), 100–119. https://doi.org/10.1207/S15327884MCA1002_2

Yamagata-Lynch, L. C. (2010). Understanding cultural historical activity theory. In *Activity systems analysis methods: Understanding complex learning environments* (pp. 13–26). Springer.

Chapter 3
Contradictions as a Methodological Tool in CHAT-Based Educational Research

Abstract This chapter positions contradictions as a rigorous methodological tool for conducting CHAT-based educational research. It opens by outlining the epistemological assumptions underpinning contradiction-focused inquiry and emphasises the need to formulate research questions that can surface systemic tensions rather than merely describe practices. The chapter then addresses research design considerations, discussing how to plan data collection in ways that enable contradictions to emerge organically across subjects, artefacts, rules and broader sociocultural contexts. A substantial section is dedicated to analytic procedures. First, traditional tools are presented, including coding strategies and the discursive manifestations of contradictions framework, which makes visible how tensions are articulated in speech, action or documentation. This is followed by an examination of how data-analysis software can support the identification, clustering and conceptualisation of contradictions, including graphical representations that illustrate relations within and between activity system nodes. The chapter also provides guidance on reporting contradiction-focused findings, including how to present resolvable contradictions and highlight unresolved contradictions as catalysts for further inquiry or institutional change. It concludes by arguing that when used methodologically, contradictions become powerful analytical devices that not only reveal systemic dynamics but also orient research towards developmental possibilities, thereby enhancing the contribution of CHAT within educational research.

Introduction

Contradictions function like cognitive dissonance, which creates a dynamic tension in a person forcing change. In so far as they lead to discomfort in a system, contradictions can also cause change. By mapping contradictions within activity systems, researchers can locate leverage points and anticipate sources of resistance or innovation. Therefore, the analysis of contradictions can be used effectively in intervention research in institutions (like schools which we well see in Chaps. 6 and 7) and organisational development.

J. Hardman, E. Rzyankina, *Contradictions in Cultural-Historical Activity Theory*, https://doi.org/10.1007/978-3-032-28650-5_3

However, the use of contradictions as methodological tools is not without its challenges. While many scholars utilise the notion of contradictions, there is some conceptual slippage in what exactly a contradiction is in practice. This is partly because a contradiction cannot be seen, which leads to difficulties when attempting to locate a contradiction or to give evidence of change that has been driven by a contradiction. In addition, contradictions do not simply arrive in a system; they emerge with a history. While systems may progress in their attempt to overcome a contradiction, they may also regress, turning away from change and resorting to historically earlier forms. An example illustrates this (and will be given in more detail in Chap. 7); while the primary object of schooling is developing students' cognitively to grasp conceptual work, what often happens in reality is that curriculum coverage becomes the object for teachers who are faced with finishing work in a limited time period. What we have here then is a primary contradiction in the object between learning and curriculum coverage. What we have seen (and this is illustrated in Chap. 7) is that rather than a progressive object of learning being the object of the activity, the activity regresses to focus on curriculum coverage (Hardman, 2023).

Yet, while contradictions are complex to deal with in a methodological manner, so too is all human activity. In fact, the difficulties of using contradictions as methodological tools represents the difficulty of studying human activity, which is messy, fluid and not linear. Considering this, CHAT's notion of contradictions is a powerful theoretical contribution, underpinning a sophisticated framework for understanding change, development and learning in human activity systems. By incorporating history, culture and the voices of all participants, CHAT enables a deep understanding of social dynamics and is applicable in a wide range of educational research.

From a Global South perspective, CHAT provides valuable methodological tools for uncovering contradictions rooted in colonial histories, economic inequalities and cultural hegemonies. For example, in educational settings, CHAT can reveal how colonial languages function as mediating tools that shape knowledge production and perpetuate inequality. It also helps illuminate the tensions that arise when indigenous knowledge systems encounter Western scientific paradigms, offering a path toward transformative engagement and decolonial futures.

Contradictions are CHAT's main methodological tool, and this chapter aims to show how they can be mobilised effectively throughout the research cycle.

Formulating a Research Question for a Study That Will Utilise Contradictions

This methodology chapter reflects the various stages of the research cycle and will act as advice/guidelines for a new researcher or who would like to find more about methodological perspective of contradiction.

A fundamental aspect of educational change is that it is riddled with contradictions (Roth & Lee, 2007). Take for example the distinction made between quantitative and qualitative research or inductive versus deductive research. Research is riddled with contradictions that one needs to navigate while driving towards and answer to your research question. In CHAT, these contradictions allow researchers to address real-life problems and adapt research questions in ways that reveal underlying causes and possible resolutions.

CHAT research is qualitative rather than quantitative because we are concerned with investigating how activities function and how contradictions arise as dynamic sites of change in and between activity systems. The solid theoretical foundation of CHAT (discussed in Chap. 1) provides a deductive rather an inductive lens through which to study actions as they unfold within activities. One may, for example (see Chap. 6) use the nodes of an activity system as a deductive framework for providing a mapping of an activity you are studying (bearing in mind that one system never works in a silo).

In order to formulate a research question that intends to focus on contradictions as sites of change, one must ask the kinds of questions that investigate change. This is especially so, for example, when one is intent on studying an intervention and its impact on an activity system. For example, the study reported on in Chap. 6 attempted to answer the following questions:

1. When a novel technology, in this case computer-based technology, is introduced into a Mathematics classroom, what pedagogical changes might arise?
2. Is there a difference between intended and enacted pedagogy with the use of computer technology?
3. Where might these changes arise and how can we track them?

The reason for posing the questions in such a way was because the research was focused on investigating change and this, in CHAT, relies on the ability to identify contradictions as they arise in the activity system. This required a detailed engagement with the data and the selection of data that could enable the depiction of an activity system and surface the contradictions in the system. As we have discussed, contradictions aren't visible; they are only visible in as discursive manifestations. One needs then, when analysing activity systems, to finely analyse language use to surface these contradictions.

In Chaps. 4 and 5 the reader can find imperial study of digital literacy of e-textbooks by engineering students and lecturers for teaching and learning process.

Primary Research Question *How do students and lecturers engage with engineering e-textbooks?*

Secondary Research Questions *What digital literacy practices do students employ while using engineering e-textbooks? (RQ1)*

- *How do lecturers deploy engineering e-textbooks in the classroom? (RQ2)*

These questions were chosen to guide the study because they enabled an integrated examination of both student and lecturer engagement with e-textbooks, thereby reflecting the dual nature of the teaching and learning process. Together, these questions aligned with the aim of understanding not only the technological adoption of e-textbooks, but also the sociocultural practices surrounding their use in engineering education.

Planning CHAT Research That Will Mobilise Contradictions

The research questions determine the choice of methodology that the researcher wishes to employ in the study (Creswell & Creswell, 2018). This methodology includes the data collection and analysis methods, as well as the theoretical underpinnings that justify their use and the nature of knowledge they yield. The research studies discussed in Part 2 all use an interpretivist and qualitative research paradigm with case-study methodology.

An interpretivist[1] research paradigm enables researchers to perceive the world from the perspectives and experiences of the participants (Thanh & Thanh, 2015). When investigating research questions, interpretive researchers rely on these experiences to construct and interpret their understanding from the data collected, rather than seeking to achieve procedural objectivity (Creswell & Creswell, 2018).

Interpretivism involves describing, understanding and explaining human action, perceptions and behaviour in a lived cultural or social context. An early proponent of this approach, the American anthropologist Clifford Geertz, argues that the task of the interpretivist is to clarify the intersubjective meanings that form the foundation of social reality in the everyday life experiences of people (1973). A later proponent, the American academic Wendy Leeds-Hurwitz, argues that researchers need to go beyond this and abstract the information that can help to answer research questions (2015). This allows researchers to uncover the 'logic' or rationality behind what might appear unfamiliar, unknown or even illogical (Cohen et al., 2017).

The qualitative research paradigm has its roots in cultural anthropology and American sociology (Kirk & Miller, 1986). Qualitative research can be characterised as intense, deeply engaging, challenging and contextualised (Cohen et al., 2017). Rather than testing hypotheses, qualitative research seeks to illuminate and understand a social phenomenon from participants' perspectives. It involves an investigative process where the researcher gradually makes sense of a social phenomenon by contrasting, comparing, replicating, cataloguing and classifying the subject of study (Miles & Huberman, 1984). The methods are often inductive in approach and involve an in-depth study of relatively few cases.

Purposive sampling is a sampling mechanism that involves intentionally selecting particular participants or settings based on their ability to offer vital information that may not be as effectively obtained through other means (Campbell et al., 2020).

---

[1] The roots of an interpretivist research paradigm are not Soviet in origin, but rather, European and derive much impetus from the work of Weber in Germany. The understanding that reality is an interpretation, however, is very much located in Vygotsky's work and that of Soviet theorists.

For example, if we are seeking to understand how a school principal views the use of generative Artificial Intelligence (AI) as a tool to teach in schools, we would purposively sample principals for this using a CHAT framework.

A case study is a research design that aims to derive meaning from a specific instance of a phenomenon, where related variables cannot be separated from the context (Merriam, 2009; Yin, 2009). The American educational psychologist Lisa Yamagata-Lynch (2010) notes that case-study approaches, with their intrinsic focus on the complexity and contextual richness of specific instances, are uniquely suited to unpacking the nuanced dynamics inherent in CHAT. Each case study presents a unique narrative, and the expected outcome is to provide specific insights and findings rather than making generalisations (Stake, 1995). Some authors (Cohen et al., 2017) believe that the results of case studies could very well be transferable to other phenomena. Case studies are also regarded as valuable supplements to other research techniques due in large part to the "thick' descriptions they provide. A 'thick descript', according to Geertz is a description that goes deeper than mere surface level and looks into interpreting what is happening in any specific context. This requires providing rich, detailed account that are embedded in the contexts in which activity happens. For this reason, case studies have proven very useful designs for CHAT researchers who seek to go beyond the surface to uncover meaning that is not immediately available to perception and is embedded in cultural life (Geertz, 1973).

An example of this can be seen in Chap. 7 where the teacher is trying to teach abstract concepts to grade 2 students. A detailed case study is used to provide a deep description of what and how each teacher in Chap. 7 teaches concepts. This enables a more realistic understanding of what is happening in each classroom than if we were, say to merely interview the teacher alone. In fact, the challenges of interviewing participants as the sole method of data collection can be seen in Chap. 6 where the difference between intended and enacted pedagogy is stark.

Triangulation is an important technique in interpretivist qualitative case-study research because it acts as a mechanism to bolster its credibility. By leveraging multiple data sources and diverse collection techniques, researchers can ensure a more robust and comprehensive understanding of their subject (Creswell & Creswell, 2018). Triangulation can be broadly categorised into three types:

- Methodological triangulation is where varied strategies are used to collect data (such as questionnaires, observations and interviews).
- Data triangulation is where information is gathered from various sources across different timeframes and settings (such as interviews as well as classroom observations that we see in Chap. 6).
- Theory triangulation is where the collected data is viewed through the lens of multiple theoretical frameworks (such as CHAT and multimodal discourse analysis).

All the research discussed in Part 2 was undertaken using both methodological and data triangulation, and additionally the research discussed in Chaps. 4 and 5 also used methodological triangulation.

As emphasised by Case and Light (2011), the relationship between the research questions and the chosen methodology is not linear, but rather dialectical in nature. This dialectical relationship mirrors the very concept of contradictions central to this study, just as contradictions emerge through the dynamic interplay between elements of an activity system, so too does research design emerge through ongoing dialogue between theoretical frameworks, empirical questions, and methodological approaches.

Gathering Data in a Way That Will Allow Contradictions to Emerge

In teaching and learning environments are shaped by a range of inherent contradictions and tensions. These emerge as students and educators navigate between traditional and evolving pedagogical practices, institutional expectations and individual capabilities, as well as formal curriculum requirements and lived educational experiences. Such tensions are not simply operational challenges but reflect deeper sociocultural processes in which tools, practices, and roles are continually negotiated. Attending to these contradictions is essential for understanding how learning actually unfolds and how educational systems respond to change.

In qualitative research, researchers immerse themselves in the study, becoming the primary data-collection instrument. The researcher's involvement in the research setting can have a valuable and positive impact rather than a detrimental one (Johri et al., 2013; Ramos, 1989). Qualitative data provides a deep understanding of the research questions where the situation is measured not by statistics and variables, but by personal experience, beliefs and values. A focus on the dialogical interaction between teacher and taught, for example, provides the space for discursive manifestations of contradictions.

Methods are tools to collect data, such as individual and focus-group interviews, observation, think-aloud protocols and online questionnaires. Together, these instruments allow the researcher to capture multiple dimensions of educational practice, including participants' experiences, interactions, and reasoning processes. Individual interviews provide access to subjective perspectives and personal meaning-making, while focus groups foreground collective sense-making and the negotiation of shared positions. Observations make visible the situated practices that participants may not articulate verbally, and think-aloud protocols generate real-time insights into cognitive and interpretive processes as they unfold. Online questionnaires complement these methods by offering broader descriptive information across a larger sample. Importantly, the combination of these methods enables the identification of contradictions, both those explicitly expressed by participants and those embedded in observed practices. By triangulating across methods, it becomes possible to detect tensions between what participants say and what they do, between individual and collective perspectives, and between institutional expectations and lived experience. In this way, methodological plurality functions not only as a strategy for data collection but also as a means of revealing contradictions that shape educational activity.

The individual interview aims to interpret the research problem through the respondent's eyes and values. Chapter 6 reports on a study of Grade 6 Mathematics lessons, and in this study, data was collected by means of an in-depth interview with the teacher, which lasted 1 h and 25 min. This interview was designed specifically to elicit the teacher's responses to the activity of teaching with technology in his Mathematics lessons and whether they perceived a shift in their pedagogy with technology. The interview schedule, provided as Table 3.1, was a methodological tool for collecting CHAT data through an interview, and it drew from the nodes of the activity system triangle.

Table 3.1 illustrate how you can deductively use the nodes of the activity system to create a framework for interviewing participants to elicit the kind of data you are looking for to populate an activity system. This is a very effective mechanism for getting a sense of participants' perceptions of actions that unfold in an activity. For data from multiple participants, however, it is more effective to use focus groups to investigate perceptions.

Focus-group interviews are a rich source of data. The strength of focus groups, according to Cohen et al. (2017), is that they encourage groups to talk about their practices. Cohen et al. (2017) also point out that when participants are relative strangers to each other, focus-group data becomes richer. The focus of a group can be what Street (2001) and Barton and Hamilton (1998) refer to as a 'literacy event', which refer to written text as integral as part of the investigation. They are concrete, observable moments that one can actually witness and describe. Examples might include a family reading a letter together, or students working through a textbook exercise. What makes these "events" is that the text isn't just present but is actively being used and matters to what's happening. When researchers focus on literacy events within groups, they are examining how literacy functions in real social contexts. This shifts attention from abstract literacy skills to how people use reading and writing together to accomplish things that matter to them. Within any group, whether a classroom, a workplace, or family, literacy events reveal the social dynamics, power relationships, values, and purposes that shape how people interact with texts. Again, this is an effective technique for understanding perceptions. However, if one's question requires that one tracks actual actions in an activity system, then observation is the more effective method to use for data collection.

Observation is a systematic technique that involves carefully selecting, observing and recording the behaviours and characteristics of social practices or phenomena (Stake, 1995). The participant observation data-collection technique (Cohen et al., 2017) allows for the collection of both linguistic data and non-linguistic data. In Chap. 6, for example, the participant observation enables the researcher to track body language. While this study is discussed in detail in Chap. 6, here we outline what a CHAT observation schedule can potentially look like for data collection.

In the study on the Grade 6 Mathematics lessons, the interview was supplemented by data based on lessons that were observed, video-taped and transcribed.

Table 3.1 CHAT interview schedule

CHAT concepts	Questions
Object	Why do you use the computer to teach Mathematics? What is the purpose of using it? How do you use it? What would you say you were working on in this lesson? Would you describe the lesson we have just observed as a fairly typical maths lesson? Do you think the computer has changed the way you teach? How? In what ways, if at all, has the use of computers assisted in your own maths teaching? In what ways has the use of the computer hindered your ability to teach in the ways in which you would like to?
Subject	How do you identify your gender? How many years have you taught? What subjects do you teach? What professional development/qualifications do you have? What is your comfort in using computer based technology? How would you say children learn? (probe for theory of learning) What, in your opinion, is the best way to teach Mathematics? Can you give an example? What factors do you think affect children's learning of maths? In what ways do you think the computer helps children to learn Grade 6 maths?
Outcome	What is produced when you use the computer software to teach Mathematics? What kind of outcomes do you hope to achieve?
Division of labour	Who does what in a lesson or activity where you use computers? What is your role? What is the student's' role? What role does technology (the computer) play?
Tools	What generative software do you use? Can you list them? What other things/materials/resources do you use in teaching? For example, PowerPoints? Notes? Readings?
Rules	What are the rules of computer use in your lessons/activities? Who decides whether an answer generated using the computer technology is correct? Does using the computer impact on pace, sequencing and selection of content in the lesson?
Transformation	Is your teaching different when using technology such as computers? How? Do students learn better using computers? Can you give a reason for your answer?

Source: adapted from Hardman, 2008. p. 69

This meant that contradictions between the teacher's intended and actual pedagogy could be identified through noting discrepancies between what was said and what was done. An observation schedule (provided as Table 3.2) was used to collect data during the observation period. Data was collected over the space of a year, with four lessons sampled per term. The sampling of a relatively few lessons was informed by the fact that pedagogy remained constant across time.

Table 3.2 CHAT observation schedule

CHAT nodes	Questions to consider during observation
Outcomes	What is the product of the lesson?
Object	What is the focus of the lesson? What is the purpose of the activity for the subject? What is the purpose of the activity for the community? What is the teacher working on? Why is the teacher working on it? What are the students working on? Why are they working on it?
Division of labour	Who does what in the lesson? (roles) Who decides what counts as meaningful? Who occupies most talk time?
Community	Who is involved in this activity? Who shares a common motive?
Rules	What are the rules of the activity? Instructional rules: Who determines pace, sequence and selection of content? Who determines the evaluative criteria? Social order rules: Who regulates the actions of the system's participants?) (discipline and communication)

Table 3.2 illustrates how one can deductively construct an observation schedule using the various nodes of the activity system to focus attention on collecting specific types of data.

Analysing Contradictions Within Data

The process of analysis involves identifying contradictions and how these contradictions relate to each other, which involves drawing together multiple insights and patterns across complementary sources of data. The process of analysis involves identifying contradictions and how these contradictions relate to each other, which involves drawing together multiple insights and patterns across complementary sources of data. Identifying contradictions requires attending to moments of tension, rupture, or misalignment in participants' accounts and observed practices, including discrepancies between stated intentions and enacted behaviours, between individual perspectives, and between institutional expectations and lived experience. Analysing these contradictions involves examining their origins, manifestations, and developmental implications, as well as tracing how they accumulate, interact, or escalate within the activity system. Through this approach, contradictions are treated not as errors or anomalies, but as analytically productive sites that reveal structural dynamics and potentials for change.

A contradiction can manifest as any of the following:

- Tensions that emerge within or between nodes of the activity system (Thuraisingam et al., 2012)

- Discursive disagreements, where participants explicitly or implicitly highlight inconsistencies, tensions or misalignments
- Conflicting motives, where different stakeholders or components of the system pull in opposite directions (Hadjistassou, 2012)
- No-win situations where participants face mutually exclusive requirements or goals
- Structural inconsistencies evident in the historical layering of the activity system, where outdated practices persist alongside new tools or expectations
- Recurring problems or disturbances that indicate deeper systemic incompatibilities
- Breakdowns or disruptions in the smooth functioning of an activity that draw attention to underlying systemic conflicts (Zhang & Luo, 2025)
- Innovations or expansive learning actions, which often signal the attempt to resolve contradictions (Lewis & Dehler, 2000; Putnam et al., 2016)

Analysing Contradictions Using Traditional Tools

While some CHAT analyses make use of computer software to analyse data, much CHAT research does not make use of this. In Chap. 6 for example one can see how the CHAT activity system framework serves as a deductive tool for data collection, while in Chap. 7 the focus on language leans more heavily towards an analysis of how contradictions emerge discursively. An example of how one can code data within a CHAT framework is given below.

Coding for Data Analysis

Table 3.3 illustrates the codes that were used to analyse the transcribed data in the study on Grade 2 lessons discussed in Chap. 7. Each sentence that could be understood out of context was coded as a unit of text. The focus on specific types of talk excluded the coding of units of text that were not relevant for the study, such as instructions of when to move and where to sit.

The codes defined in Table 3.3 allow the researcher to count and quantify how particular talk is used in a lesson which can assist the qualitative interpretation of this data. Table 3.4 illustrates the percentage of talk per category coded in the observed lessons, report in Chap. 7. This allows us to see how much, as a percentage of overall talk, is focused on the specific talk that the researcher is interested in, providing a broad view of what is happening in the lesson, before a more refined and nuanced qualitative analysis is undertaken.

Table 3.3 Analytical codes used in the study of pedagogy in Grade 2 Life Skills lessons

Code	Definition	Example
ST	Science term, not elaborated	'Cats are predators'
ESC	Elaborated scientific (abstract) concept	'They have retractable claws to hunt very quietly and to protect themselves. And so their bones are different to ours.'
ED	Everyday concept	'My cats eat grass.'
CQ	Closed question	'So small house cats eat?'
OQ	Open question	'Can you explain what the word 'predator' means?'
FE	Feedback elaborated	'So, they are predators, they are hunters.'
FN	Feedback not elaborated	'Yes, that's right.'

Table 3.4 Coding of teacher talk across a lesson in the study of pedagogy in Grade 2 Life Skills lessons

Code	Mrs Jenkins %	Mrs Smit %
Scientific term	8	21
Elaborated scientific concept	16	10
Everyday concept	17	8
Closed question	18	24
Open question	2	–
Feedback elaborated	22	11
Feedback not elaborated	17	26
TOTAL	100%	

The Discursive Manifestations of Contradictions Framework

Quantifying talk as above, enables the researcher to see where potential contradictions can arise discursively in the data. Of course, this is a static picture of a percentage of talk and cannot enable a deeper interpretation of a contradiction and what it achieves. What the data in Table 3.4 do allow though, is a brief static view of where we may potentially be looking for contradictions. This data also serves as a comparative snapshot of how language is used by different teachers in the same grade teaching the same topic. This is discussed in further depth in Chap. 7.

To address the complexity of contradictions and their discursive phenomena, Engeström and Sannino (2011) developed an analytical framework known as the discursive manifestations of contradictions. This framework helps researchers understand the practices associated with contradictions, how to deal with them, and how to resolve them.

Within this framework, there are four types of manifested contradictions: dilemmas, conflicts, critical conflicts and double binds. These different types of contradictions provide researchers with a structured approach to analysing and interpreting the way in which contradictions manifest in participants' language. Table 3.5 summarises the features of each type of contradiction, shows how such a contradiction will be resolved, and gives the linguistic cues that will help researchers identify it.

Table 3.5 The four types of contradiction in Engeström and Sannino's discursive manifestations of contradictions framework

Manifestation of contradiction	Features	Resolution	Linguistic cues
Dilemma	Expression of incompatible evaluations	Denial	'on the one hand ... on the other hand', 'yes, but', 'I didn't mean that', 'I actually meant'
Conflict	Arguing, criticising	Compromise, submitting to authority or majority	'no', 'I disagree', 'this is not true', 'yes', 'this I can accept'
Critical conflict	Facing contradictory motives in social interaction, feeling violated, guilty or helpless	Finding new personal direction and negotiating a new meaning	'I now realise that…', personal, emotional and moral accounts, narrative structure, vivid metaphors
Double bind	Facing pressing and equally unacceptable alternatives in an activity system	Practical transformation (going beyond words)	'we', 'us', 'we must', 'we have to', 'let us do that', 'we will make it', pressing rhetorical questions, expressions of helplessness

Source: Adapted from Engeström and Sannino (2011, p. 375)

Analysing Contradictions Using Data-Analysis Software

Computer-aided qualitative data-analysis software (CAQDAS) facilitates the systematic organisation, coding, retrieval and comparison of qualitative data, while also improving transparency and traceability in the analytic process. There is a variety of proprietary software (e.g. ATLAS.ti, NVivo, MAXQDA) and free/open-source software (e.g. QualCoder, RQDA, Taguette) available.

Using one of these as an analytical tool has a number of advantages. Firstly, it allows the researcher to quickly and comprehensively search within the data. This is because the software can handle multiple and overlapping codes without losing the context of analysis. Secondly, codes can be combined using the filter function, which is useful for identifying more detailed connections and relationships. Thirdly, it can be used to give others access to the data and coding, allowing them to see the units of meaning and relationships within the data analysis.

The software ATLAS.ti also supports qualitative researchers in theory building, by allowing researchers to connect codes. This helps researchers to develop higher-order classifications and categories. Furthermore, this software helps researchers formulate propositions that align with their data. The research study that is discussed in Chaps. 4 and 5 was undertaken using ATLAS.ti.

In addition to this, the discursive manifestations of contradictions framework can be used alongside CAQDAS as a structured analytic tool. Discursive manifestations (such as dilemmas, conflicts, double binds, and critical conflicts) can be operationalised as coding categories within the software, enabling systematic tagging of

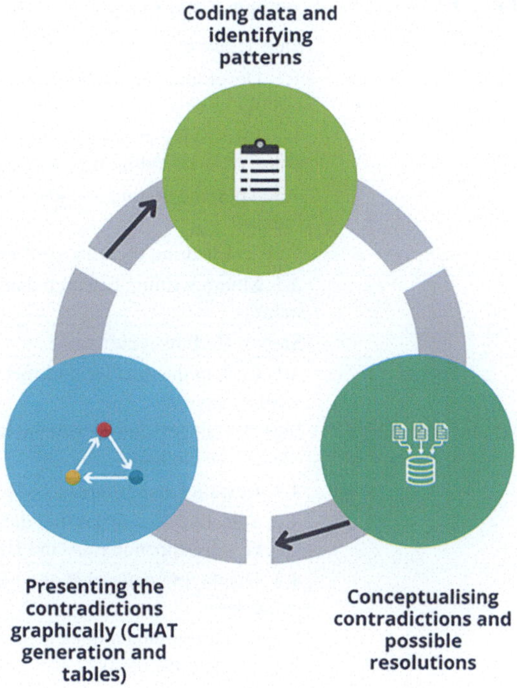

Fig. 3.1 A cycle diagram (cyclical process) of contradictions using CAQDAS

utterances or segments that signal tension within the activity system. CAQDAS facilitates this process by allowing the researcher to organise, compare, and cluster coded excerpts across different participants and data sources, while the framework provides the conceptual lens through which these coded segments are interpreted. Used together, the framework and CAQDAS enhance analytic rigour: the software supports transparency and traceability of coding decisions, and the framework guides the identification, classification, and theorisation of contradictions as they emerge in discourse and practice.

Figure 3.1 shows how data analysis of contradictions can done using CAQDAS, which is discussed in more detail below.

Coding for Data Analysis

The coding process can be divided into two phases. The first phase focuses on describing the data and establishing a coding system. The second phase focuses on querying the data to find answers and identify relationships aligning with the research questions. Table 3.6 below shows the first phase (steps 1–3) and the second phase (steps 4–6), which will be discussed in detail below.

The first phase is a descriptive analysis, and the goal in this phase is to thoroughly explore the data. This entails reading through transcripts and field notes, and

Table 3.6 Six steps of coding for data analysis

Phases	Step 1: Inspection cycle
Descriptive analysis (inductive coding)	1.1. Developing a codebook (see Appendix A) and exporting this to the CAQDAS (e.g. ATLAS.ti, Nvivo) (see Tables 3.7 and 3.8)
	1.2. Identifying codes by using an open coding strategy and the initial phase of auto-coding (in vivo coding) by examining the entire data set
	1.3. Applying coding and identifying families of codes and possible themes
	Step 2: Ensuring the trustworthiness of the coding process
	2.1. Memo writing: research diary, methodological memos and analytical memos
	Step 3: Coding cycle
	3.1. Coding the data set multiple times, merging and splitting codes, and writing memos
Thematic synthesis and interpretation	Step 4: Categorisation cycle includes selective and focus coding (deductive coding)
	4.1. Revising and grouping codes to elaborate possible categories
	4.2. Selective coding: coding data from a theoretical perspective of CHAT (see Appendices A and B).
	4.3. Displaying networks[a] of codes and displaying relationships among categories
	Step 5: Developing a thick description of data in narrative format (interpreting the data)
	5.1. Examining data and identifying stories within the data, using the search button for linguistic cues (as described in Table 3.1) for further analysis of the discursive manifestations of contradictions
	5.2. Writing narratives
	Step 6: Conducting an activity system analysis (explaining data)
	6.1. Drafting activity systems based on narratives
	6.2. Identifying and reporting the final findings

[a]A network refers to a visual representation of data consisting of nodes or points interconnected by links or lines. These networks are used to display participant actions, events and processes in a coherent and connected manner

listening to audio files, in order to identify noteworthy patterns and trends; as well as generating word clouds and word lists, conducting searches based on CHAT concepts and coding the data.

In Step 1: Inspection cycle, the aim is to establish a comprehensive code (inductive coding entails generating analytical codes directly from the data rather than applying a pre-existing coding book) list that describes the various issues, aspects, phenomena and themes present in the data. Through this process, these elements are named and examined for similarities and differences, resulting in a structured code list. The code list is refined, and additional cycles of observation and data collection are conducted, to ensure comprehensive coding and the development of a complete coding schema. The researcher assigns comments to each of these codes, to indicate the exact meaning of the code and to help avoid bias in the coding process. The researcher should also create memos and reflect on the process of data analysis. This allowing the researcher to keep track of the process of data analysis and better understand the tensions in the activity system.

In Step 2: Ensuring the trustworthiness of the coding process, the researcher writes research diary, memos (place for reflection), marking the segments that are of interest. The research diary can be used throughout the project to record procedural decisions, fieldwork reflections, practical challenges and contextual notes arising during interviews and transcription. Methodological memos will capture decisions related to the research design and analytic strategy, including how codes are developed, how categories are refined, and how methodological adjustments are justified over time.

In Step 3: Coding cycle (inductive coding), the aim is to query the data to find answers and identify relationships aligning with the research questions. In addition to inductive coding, the analytic process involves systematic refinement of the code system, including the merging and splitting of codes as patterns become clearer. CAQDAS tools are used to query the data based on the previous coding. These queries yield results that can be presented in numerical form, coded quotations or visualisations. However, the actual analysis process occurs during the writing, where the researcher summarises and interprets the outcomes. Use of the memo function can facilitate this process. Through writing comments and memos, the researcher advances the analysis step by step, delving deeper into the data, examining details, gaining a better understanding of how elements are connected, and uncovering intricate relationships within the data.

In Step 4, the Categorisation cycle, the aim is to move from descriptive labels toward more structured analytical groupings. This begins with generating codes inductively through a close reading of the data, allowing concepts, distinctions and meanings to emerge from participants' accounts rather than imposing a preconceived structure. Once an initial inductive code set has been developed, the researcher organises the codes into folders, categories and sub-codes. Deductive coding may then follow, using theory to inform higher-level classifications and to examine how empirical material speaks to predefined concepts. Throughout this stage, the researcher creates links between coded segments and constructs network views to visualise relationships and dependencies within the data. Once the data is classified, the researcher makes formal and informal connections, identifies patterns, and investigates regularities and variations. This provides a meaningful conceptual foundation for subsequent interpretations. In addition, the researcher examines the concept of contradiction and explores its discursive manifestations in the data, attending to how tensions are articulated, negotiated or normalised in participants' narratives.

In Step 5: Developing a thick description of the data in narrative format, the researcher begins to link the inductively generated codes to the theoretical framework, analysing patterns, relationships and meanings across cases and writing the analytic narratives. In doing so, the researcher interprets how the empirical material speaks to key concepts, mechanisms or tensions within the chosen theory. Once this interpretive stage is complete, the researcher creates an analytical framework for the deductive code analysis that will follow, which is presented in Table 3.4. This framework enables the transition from open inductive exploration to theory-driven examination, supporting analytical coherence and ensuring that subsequent deductive coding is grounded in both empirical material and theoretical commitments.

In Step 6: Conducting an activity system analysis, the researcher maps the empirical material onto the elements of the activity system in order to identify where tensions and contradictions emerge, how they manifest, and what systemic relations are implicated. This step involves examining interactions between subjects, tools, rules, community, division of labour and object, and then charting how these relations produce breakdowns, conflicts or misalignments. The analysis is supported by evidentiary excerpts from the data and may be summarised through tables that classify the level of contradiction, indicate where it appears in the system, describe its manifestation, present empirical evidence, and consider how the activity system resolves, transforms or could be deliberately intervened in to address the underlying tension.

Conceptualising Contradictions

When using the discursive manifestations of contradictions framework, in order to understand contradictions and their possible resolution the researcher investigates the four manifestations of contradictions presented in Table 3.1, which are dilemmas, conflicts, critical conflicts and double binds (see Appendix B). A codebook is created during qualitative analysis to organise, define, and manage the codes that are applied to the data. In an inductive approach, the process typically begins with close reading of the transcripts and assigning provisional labels to meaningful segments of text.

The researcher also identifies the principal primary contradiction of the activity system. Within this principal contradiction the researcher must also identify which aspect of the contradiction is currently dominant.

Graphical Representation of Contradictions

Manifested contradictions can be graphically represented using the CHAT model of a single activity system, with each stakeholder and tool placed within the relevant sections of the CHAT triangle (see Fig. 3.2, which relates to the discussion in Chap. 5). This allows for a closer inspection of the dynamics of the activity system. Graphical representations can also be used for intersecting activity systems (see Fig. 5.1 in Chap. 5), which illustrate interactions and contradictions between these systems.

Depends on the qualitative software the researchers can use the network representations which help to demonstrate initial and work in progress and final coding and findings of the study. Network offer an organised and coherent depiction of the relationships and connections within the data, resulting in a clearer and more comprehensive comprehension of the research subject. Utilising network visualisations

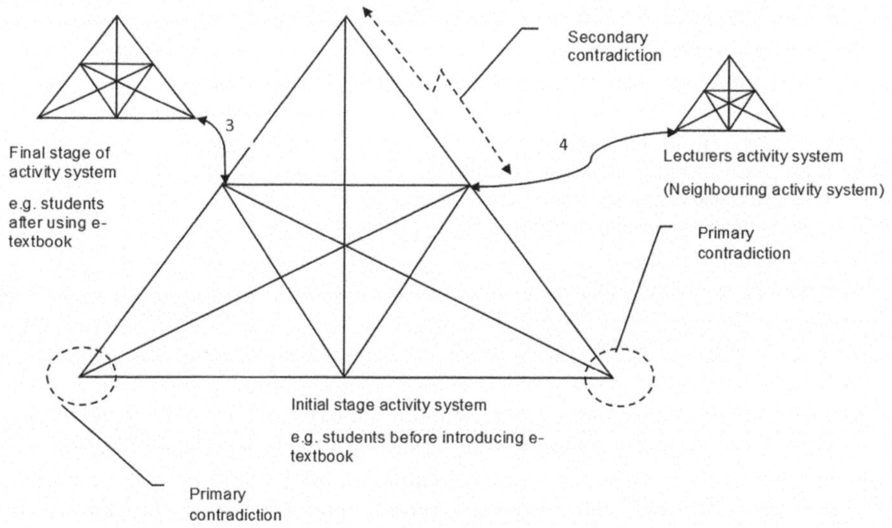

Symbol	Description
Primary contradiction	This contradiction appear within elements
Secondary contradiction	This contradiction appear between elements of the system
3 Tertiary contradiction	This contradiction is between different versions of the same activity system. e.g. before and after or old and new model of teaching
4 Quaternary contradiction	This contradiction is between neighbouring activity system e.g. students – parents or students first year – students second year

Fig. 3.2 A graphical representation of contradictions in the research on students' usage of e-textbooks

contributed to a more profound and meaningful interpretation of the data, promoting effective communication and discussion of the research outcomes.

In Table 3.7 below, can be used as structured analytic tools for organising complex findings about contradictions within an activity system and for making the analytic process visible to the reader. In Cultural-Historical Activity Theory, contradictions are not simply problems or conflicts; they are systemic tensions that manifest in practice and discourse and that can drive change within the activity system. The table format supports this by breaking the analysis into intelligible components.

At a classificatory level, the table helps identify the level of contradiction (primary, secondary, tertiary or quaternary) and the location within the activity system where it emerges (for example in the subject, tools, rules or community). This prevents analytical flattening in which all tensions are treated as equivalent, and instead

Table 3.7 Questions and concepts for deductive coding in a CHAT analysis

Questions to ask when analysing data	CHAT concepts
What is produced in this episode (or transcript)?	Outcome
What tools are used?	Mediating artefact
What is the object/focus of this episode? What is the purpose of the activity for the subject? What is the participant (student and/or lecturer) working on? Why is this participant working on this?	Object
Who does what in this episode? Who determines what is meaningful?	Division of labour
What community is involved in this episode? What group of people work together on the object?	Community

makes the structural character of the contradiction explicit. The table then encourages interpretive depth by requiring a description of the type of manifestation and its underlying cause. This step draws attention to how contradictions surface in practice, such as through stalled processes, conflicting expectations, breakdowns in communication, or competing values, and what systemic mechanisms give rise to these tensions.

The inclusion of empirical evidence is critical because it grounds the analysis in the data rather than allowing the researcher to make purely conceptual claims. Evidence typically takes the form of quotations from interviews, extracts from observational notes, or documentary material.

The table also supports developmental analysis. Columns focused on resolution highlight whether and how the contradiction was resolved within the activity system or whether it remains unresolved. CHAT treats contradictions as sources of development; therefore, the resolution column can also point towards how the system underwent transformation or how it could transform in the future. In cases where contradictions have not been resolved in practice, the table format offers space to propose possible interventions that could alleviate or reconfigure the systemic tension. These might involve retooling, changes to rules or policies, redesign of roles, or shifts in the division of labour. This makes the table not merely descriptive but generative, by linking analysis to developmental possibilities.

Reporting on Contradiction-Focused Research

In educational research, the goal of contradiction analysis is to enable and promote the possibility of educational change. However, even if this change occurs, it does not necessarily eliminate tensions. Because the system is constantly changing, these specific tensions may persist or they may evolve into new tensions.

Resolving Resolvable Contradictions (See Table 3.8)

After discussing contradictions as part of the research findings, it is important to address possible resolutions of those contradictions as part of the study's contribution. In Cultural-Historical Activity Theory, contradictions are not merely diagnostic categories; they are developmental drivers that point toward opportunities for change and improvement within the activity system. By analysing how a contradiction might be resolved, the researcher demonstrates how empirical insights can inform practical, pedagogical or organisational interventions. This may include clarifying which elements of the system could be reconfigured (for example, tools, rules, division of labour, or object), identifying enabling conditions for change, and outlining the actors responsible for initiating or sustaining such developments. Integrating possible resolutions into the discussion therefore strengthens the contribution of the research by translating theoretical insight into actionable knowledge, making the findings useful for design, policy, and future practice.

Table 3.8 Contradiction analysis of the data

Level of contradiction	Where it manifests in the activity system	Type of manifestation and cause	Particular manifestation	Evidence	Resolution of the contradiction
What level of contradiction is it? E.g. primary level	Who is the subject of the activity? Is it within an element or between elements or activities? E.g. students' activity, lecturers' activity	As what type of manifestation does the contradiction appear? E.g. double bind What causes this type of contradiction?	What does it manifest as? As example, *reduced engagement*, where students skim rather than read in depth, resulting in shallow comprehension or surface learning.	What quotations from the data support this? *Include participants quotes here from your study.*	How could this be resolved? What could be transformed? What intervention could be undertaken? *Propose possible resolution.* In context of Chaps. 4 and 5 students suggested to provide full training on how to operate the e-textbook.

Raising Unresolved Contradictions for Further Investigation

In cases where contradictions remain unresolved, the researcher should make these visible and explicitly flag them as areas requiring further inquiry rather than forcing premature conclusions. This involves clearly articulating the nature of the unresolved tension, the parts of the activity system involved, and why resolution could not be identified within the scope of the present study. By doing so, the researcher contributes to the cumulative development of knowledge by delineating gaps that future researchers can address. These flagged issues may indicate emerging phenomena, structural constraints, theoretical blind spots, or practical barriers that require additional empirical work, methodological innovation, or longitudinal tracking. Documenting unresolved contradictions in this way invites other scholars to pursue solutions, test interventions, or reconceptualise the problem space, thereby extending the research agenda beyond the boundaries of a single study.

Conclusion

Contradictions are the main methodological tool in CHAT and can be used to trace the dynamic interplay between stability and change within activity systems. In qualitative interpretivist research, it is sometimes said that "the more things change, the more they stay the same," in the sense that new software tools and analytic techniques often continue to serve long-standing scholarly goals such as sense-making, interpretation and the construction of meaning from situated data, even as their forms become more technologically sophisticated. In this respect, contemporary qualitative tools such as CAQDAS may accelerate certain tasks or enable new forms of visualisation, yet the underlying interpretive cycle of reading, coding, comparing, memoing, theorising and returning to the data remains remarkably consistent. This continuity highlights an important feature of CHAT: although tools mediate activity and can transform the conditions of learning, adaptation and development, they do not fundamentally erase the need for human interpretive labour; rather, they reshape how such labour is organised, distributed and made visible.

At the same time, contradictions make it clear that new tools are not neutral replacements for old ones. They introduce new affordances and constraints, alter the division of labour, and reposition actors within the system in ways that can produce both continuity and discontinuity. Thus, contradictions serve not only as markers of disruption but also as methodological signposts for analysing how technologies intersect with the research cycle, pedagogy or organisational practices. In CHAT-informed interpretive work, analysing contradictions enables the researcher to map how established goals persist, how new tensions arise, and how the system as a whole accommodates or resists transformation.

In this book, we draw on examples in STEM education to illustrate this usefulness. We show how university students studying engineering courses navigate and engage with new tools such as e-textbooks in the curriculum. We also use this framework to explore Mathematics and Life Skills education in schools. These examples demonstrate that CHAT offers both a robust and widely applicable analytical approach for studying educational practices and developmental processes across diverse sites. It sheds light on how learners, teachers and institutions confront tensions brought about by digitalisation, shifting curricula and evolving expectations, and how these tensions become drivers of change rather than merely obstacles.

A final reflection is warranted regarding the Global South and 4th-Generation CHAT. Although this chapter has foregrounded methodological concerns, the broader CHAT literature increasingly recognises the importance of attending to cultural, geopolitical and historical contexts, particularly in the Global South where educational ecologies, resource distributions, linguistic diversity and postcolonial dynamics shape activity systems in distinctive ways. To the extent that 4th-Generation CHAT extends the unit of analysis to networks, ecosystems and translocal relations, it also offers a productive lens for capturing how STEM education in the Global South is embedded in global flows of knowledge, technology and policy. If future chapters engage more explicitly with these dialectical scales, local practice, institutional context, and transnational structures, then contradictions can serve as a methodological bridge: connecting micro-level experiences with broader systemic forces, and situating educational development in contexts where "the more things change, the more they stay the same" may reflect both resilience and structural constraint.

References

Barton, D., & Hamilton, M. (1998). *Local literacies: Reading and writing in one community*. Routledge.

Campbell, S., Greenwood, M., Prior, S., Shearer, T., Walkem, K., Young, S., Bywaters, D., & Walker, K. (2020). Purposive sampling: Complex or simple? Research case examples. *Journal of Research in Nursing, 25*(8), 652–661. https://doi.org/10.1177/1744987120927206

Case, J. M., & Light, G. (2011). Emerging research methodologies in engineering education research. *Journal of Engineering Education, 100*(1), 186–210.

Cohen, L., Manion, L., & Morrison, K. (2017). *Research methods in education* (8th ed.). Routledge. https://doi.org/10.4324/9781315456539

Creswell, J. W., & Creswell, J. D. (2018). *Research design: Qualitative, quantitative, and mixed methods approaches* (5th ed.). SAGE Publications.

Engeström, Y., & Sannino, A. (2011). Discursive manifestations of contradictions in organizational change efforts: A methodological framework. *Journal of Organizational Change Management, 24*(3), 368–387.

Geertz, C. (1973). Thick description: Toward an interpretive theory of culture. In *The interpretation of cultures: Selected essays* (pp. 3–30). Basic Books.

Hadjistassou, S. K. (2012). An activity theory exegesis on conflict and contradictions in networked discussions and feedback exchanges. *CALICO Journal, 29*(2), 367–388.

Hardman, J. (2008). Researching pedagogy: An activity theory approach. *Journal of Education, 45*(1), 65–95.

Hardman, J. (2023). *A cultural-historical approach towards pedagogical transitions: Transitions in post-apartheid South Africa*. Bloomsbury Academic.
Johri, A., Roth, W. M., & Olds, B. M. (2013). The role of representations in engineering practices: Taking a turn towards inscriptions. *Journal of Engineering Education, 102*(1), 2.
Kirk, J., & Miller, M. L. (1986). *Reliability and validity in qualitative research*. Sage.
Leeds-Hurwitz, W. (2015). Thick description. In K. Tracy, C. Ilie, & T. L. Sandel (Eds.), *The international encyclopedia of language and social interaction* (Vol. 3, pp. 1515–1520). Wiley-Blackwell. https://doi.org/10.1002/9781118611463.wbielsi072
Lewis, M. W., & Dehler, G. E. (2000). Learning through paradox: A pedagogical strategy for exploring contradictions and complexity. *Journal of Management Education, 24*(6), 708–725.
Merriam, S. B. (2009). *Qualitative research: A guide to design and implementation*. Jossey-Bass.
Miles, M. B., & Huberman, A. M. (1984). Drawing valid meaning from qualitative data: Toward a shared craft. *Educational Researcher, 13*(5), 20–30. https://doi.org/10.3102/0013189X013005020
Putnam, L. L., Fairhurst, G. T., & Banghart, S. (2016). Contradictions, dialectics, and paradoxes in organizations: A constitutive approach. *Academy of Management Annals, 10*(1), 65–171.
Ramos, M. C. (1989). Some ethical implications of qualitative research. *Research in Nursing & Health, 12*(1), 57–63.
Roth, W.-M., & Lee, Y.-J. (2007). "Vygotsky's neglected legacy": Cultural-historical activity theory. *Review of Educational Research, 77*(2), 186–232. https://doi.org/10.3102/0034654306298273
Stake, R. E. (1995). *The art of case study research*. Sage.
Street, B. V. (2001). Literacy empowerment in developing societies. In *Literacy and motivation* (pp. 273–281). Routledge.
Thanh, N. C., & Thanh, T. T. (2015). The interconnection between interpretivist paradigm and qualitative methods in education. *American Journal of Educational Science, 1*(2), 24–27.
Thuraisingam, T., Kaur, P., Yeo, S., Briguglio, C., Sanderson, G., Mahmud, S., & Wallace, M. (2012). An activity theory approach to fair assessment moderation in transnational education. *Journal of Interdisciplinary Research in Education, 2*(1), 1–15.
Yamagata-Lynch, L. C. (2010). *Activity systems analysis methods: Understanding complex learning environments*. Springer Science & Business Media.
Yin, R. K. (2009). *Case study research: Design and methods* (4th ed.). Sage.
Zhang, L., & Luo, T. (2025). 'I've never experienced learning like this before': Examining a Chinese doctoral student's expansive learning from the activity theory perspective. *Studies in Continuing Education*, 1–21.

Part II
Case Studies Using Contradictions in Educational Research

Chapter 4
When the Mediating Artefact Disrupts the Original Object: How e-Textbooks Impact Engineering Students' Learning of Physics

Abstract This chapter examines how the adoption of e-textbooks as mediating artefacts shapes Engineering students' learning of Physics within a university context. Drawing on Cultural-Historical Activity Theory (CHAT), the chapter analyses the students' activity system to uncover systemic tensions produced when traditional pedagogical practices intersect with digital learning tools. Findings indicate that the mediating artefact itself can disrupt, distort or reorient the original learning object rather than simply facilitate it. Five key contradictions emerge from the empirical data. First, students grapple with the dual task of learning Physics content while simultaneously learning to operate the digital platform. Second, limited functional engagement contrasts with the effort required to explore e-textbook affordances. Third, a desire for unrestricted access is challenged by institutional licensing constraints. Fourth, individual study preferences conflict with collaborative affordances embedded in e-textbook platforms. Finally, students navigate between pre-curated, institutionally approved content and diverse yet unreliable online resources. Collectively, these contradictions demonstrate that the mere provision of digital learning materials does not guarantee epistemological access. Instead, the mediation of learning becomes contingent on institutional policies, technological infrastructures, and students' digital literacies. The chapter concludes by considering how educational design and institutional decision-making may better align the mediating artefact with the intended learning object.

Introduction

Textbooks are a mediating artefact used by students to achieve their object of understanding the content and skills covered in a particular course. When e-textbooks replaced paper textbooks at a South African university of technology, the tensions that resulted could not be directly observed, but could be discerned in the way in which students adapted or resisted the new digital literacy practices that were required. A CHAT analysis helps to reveal why students responded as they did to

J. Hardman, E. Rzyankina, *Contradictions in Cultural-Historical Activity Theory*, https://doi.org/10.1007/978-3-032-28650-5_4

this new digital tool, one which was meant to assist them to learn but which became something more that they had to learn.

A CHAT Analysis of the Students' Activity System

As shown in Fig. 1.3 in Chap. 1, an activity system can be represented in the form of an activity triangle containing intersecting lines. In the top half of the triangle, the mediating artefact is used by the subject to achieve the object of the activity. In the bottom half of the triangle, the rules, community and division of labour affect how this happens.

The mediating artefact is made up of the group of resources that the subject draws on during the activity. In this activity, a significant tool within the mediating artefact is the Physics e-textbook, which has been incorporated into the learning management system (LMS) of the university of technology (this university used under LMS Blackboard). Language and mathematics are semiotic tools that also make up part of the mediating artefact, both as part of the e-textbook and as part of the lectures and tutorials that the students attend.

The subject is the group of Chemical Engineering or Maritime Science students who were instructed to use the e-textbook as part of their first-year Physics course in 2021. Physics is one of the courses with a high failure rate. The students in this activity were part of the extended curriculum programme (ECP) that supports disadvantaged students who did not meet the minimum entrance requirements for their degrees.

The object is the goal (motive for collective goal) of the activity. In this study, the object is learning Physics. In educational activity for some students the object might be to gain a deep conceptual understanding of the course, while for others the object might be to understand and remember enough to pass an exam.

The rules refer to the formal policies or informal regulations and norms of the activity system. These rules affect how the activity takes place at the university of technology. The rules from these two departments were students should use e-textbook as prescribed course material. The e-textbook was available for all registered students at LMS Blackboard. The paper book was also available at the library.

The community is the social group that the subject belongs to while engaging in the activity. In this activity system, the community includes the subject's peers, lecturers, information technology (IT) system administrators, librarians and staff of the Centre for Innovative Educational Technology (CIET), as well as the developers of the e-textbook—in other words, all the people who support the subject's pursuit of the object.

Division of labour refers to how tasks are shared among these members of the community, which is affected by the unequal power dynamics in the activity system. In this activity system, the subject has less power than the lecturers within the community.

While mediated action occurs between the subject, mediating artefact and object at the top of the triangle, social practice is concerned with the nodes at the bottom of the triangle: rules, community and division of labour. E-textbooks, as a mediating artefact, are mediate interaction while reading, as a social practice, follows patterns of a specific context of an activity system.

How Contradictions Manifest When Students Are Required to Use e-Textbooks

The tensions that emerge from students' use of e-textbooks are visible in the data set. The contradictions that underly these can be detected by examining how they manifest in practical situations. I used the discursive manifestations of contradictions framework proposed by Engeström and Sannino (2011) to analyse how four distinct manifestations of contradictions emerge from the way people speak about their activities.

These contradictions are shown using the letters A–E in Fig. 4.1, which depicts a previous activity system of these students and the new activity system of these students. All the contradictions implicate various nodes of the new activity system. However, for analytical purposes, in the discussion that follows I highlight only those nodes that are most involved in the tensions, that most emphasise the digital literacy practices, and that most clearly open up possibilities for educational development through the use of e-textbooks.

In order to identify these contradictions, I employed the concept of linguistic cues, as suggested by Engeström and Sannino (2011), during the preliminary data-analysis phase. However, in the next stage of data analysis, specifically focusing on the student participants of this study, not all manifestations of contradictions were consistently supported by linguistic cues. The linguistic cues became analytically meaningful in this study only after the students' utterances had been examined in relation to their contextual meaning and the quotations had been interpreted accordingly. Rather than explicitly using the exact linguistic cues identified by Engeström and Sannino (2011), students articulated their experiences in more implicit and situated ways. As a result, the linguistic cues listed in Engeström and Sannino's framework could not be applied directly; instead, their meanings had to be analytically inferred and mapped onto each identified contradiction.

In the following sections, I discuss the contradictions presented in Fig. 4.1 in greater detail.

Contradiction A: Learning Physics with e-Textbooks Versus Learning How to Operate e-Textbooks

The introduction of e-textbooks in the learning process has created a contradiction in the primary object of the activity system, which is learning Physics (see A in Fig. 4.1). The intended goal of the students is to learn through the use of e-textbooks, but a contradiction arises as the students find themselves primarily focused on

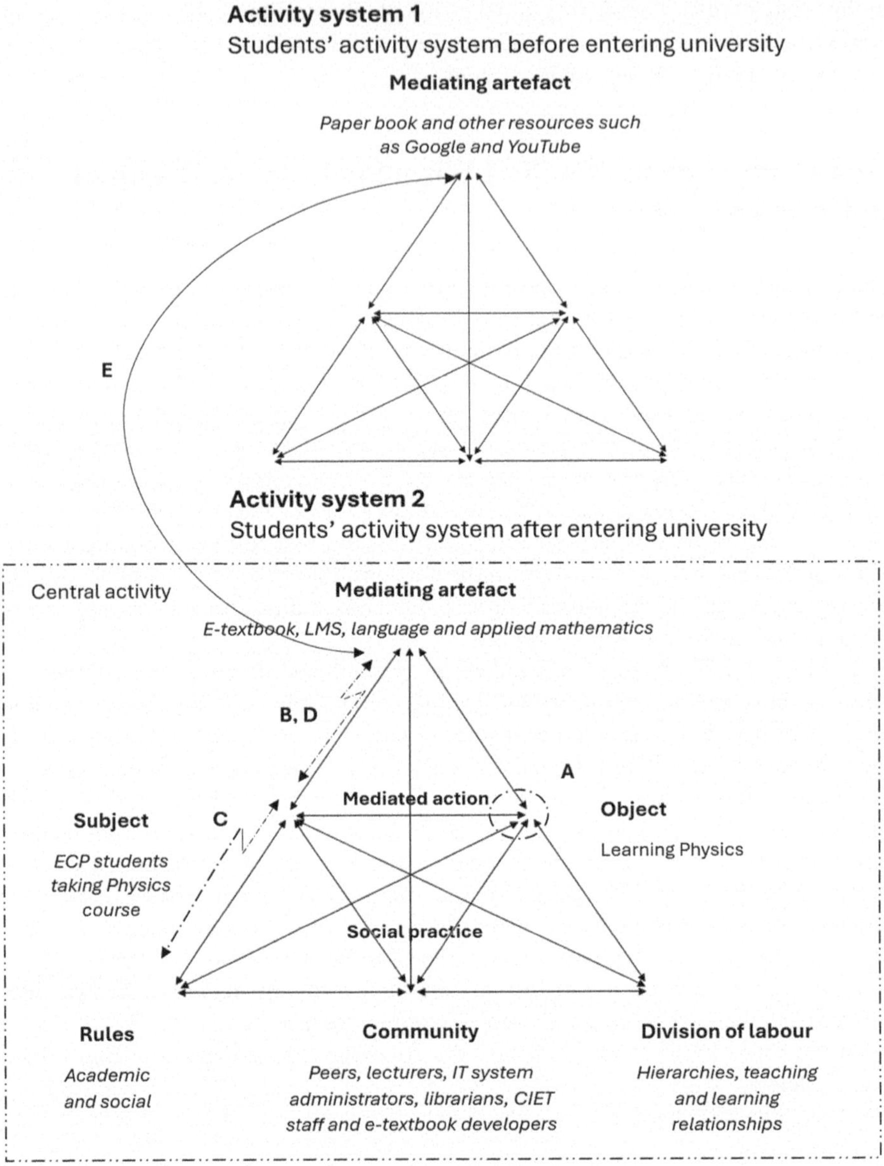

Fig. 4.1 Five contradictions in the students' activity system

Contradiction (presented in terms of its contradictory aspects)	**Type and location of contradiction**	**Type of contradictions**
A: Learning Physics with e-textbooks versus learning how to operate the e-textbooks	Primary contradiction within the object of the student activity system	Dilemma Resolution: provide extensive training on introduction of e-textbook/include in curriculum digital or AI literacy
B: Limited usage of e-textbooks versus investing energy to explore the functionalities of e-textbooks	Secondary contradiction between the subject and the mediating artefact	Dilemma Resolution: provide extensive training on introduction of e-textbook
C: Desire for unlimited access versus the restricted distribution of e-textbooks	Secondary contradiction between the subject and the rules	Conflict Resolution: create platform open educational resources in future for students
D: Preferences for solitary study versus the collaborative study affordances of e-textbooks	Secondary contradiction between the subject and the mediating artefact	Conflict Resolution: the intentional promotion of collaborative learning from the first day of the study. By explicitly demonstrating the benefits of collaboration and embedding collaborative practices into learning activities, students are supported in developing shared responsibility for learning. The e-textbook functions as a mediating artefact in this process, enabling collaborative engagement with content, peer discussion, and the co-construction of knowledge.
E: Pre-curated information in e-textbooks versus unreliable online resources	Tertiary contradiction between the mediating artefact in the previous version of the activity system and the mediating artefact in the new version of the activity system	Conflict Resolution: Training on curation and fact checking.

Fig. 4.1 (continued)

learning literacy skills associated with the e-textbooks rather than the actual Physics content.

This primary contradiction manifests as a dilemma. A dilemma is when individuals face two conflicting paths, both of which are fraught with uncertainty. Here, students grapple simultaneously with the dual responsibilities of mastering intricate Physics concepts and becoming adept at using e-textbooks. This dichotomy complicates their learning journey, pulling their focus in two directions.

Of the student respondents, 3 students did not know what an e-textbook was, 26 were unsure, and 44 confirmed they were familiar with e-textbooks. However, while some students may have had a limited understanding of what e-textbooks were, this did not mean they possessed the digital literacy skills needed to study effectively using them.

Student 3 predicted, 'Students will end up quitting e-textbooks, even if it is a lot of opportunities for study. But because they don't even know how to use it and won't even try.' In contrast, Student 14 revealed, 'I don't know how to do so but I would love to know how [to use e-textbooks]'. Both these students expressed need for explanation and practices on how to study with e-textbook.

Student 12 explained their initial experience of an e-textbook, remembering, 'On the first day, whenever we started the course, we couldn't open it. I didn't know when to open it or how to use it.' Students also struggled at the beginning of the semester because they could not find the e-textbook in the LMS. Some students sought alternative access to the prescribed textbook in PDF format. However, the tension remained. Student 12 expressed their frustration by saying, 'I don't have the books that I am going to use. Like, just since every textbook that you're gonna use [is] a PDF [sic].'

This indicates that some students are unfamiliar with digital texts and tools, and that they require assistance to access the digital text. The students also need better instruction on how to use e-textbook features effectively. Lecturer 2 emphasised the need for clear instructions and explanatory videos to help students navigate and utilise the new resources:

> E-textbooks got some very nice things you can do there. However, I don't see any way to get instructions on how to access the videos. How do you select a particular chapter to practise on? What do you do if, like, there are a lot of features on there that are brilliant? But for someone who doesn't even know what the feature is, for instance, assuming there's a student who doesn't know how to use a highlighter, just having that highlighter thing on there without having something that explains what it does makes it kind of redundant. Many students won't use it because they don't know what it can do.

The primary contradiction in the object of the students' activity is also evident in the response of Lecturer 2:

> Students are supposed to learn Physics, but now they must learn how to download and operate the e-textbook. And they must learn how to do this and other things. So, it's a learning curve of getting comfortable with the whole thing and readjusting to using an e-textbook compared to a physical one.

This shift in learning priorities is encapsulated by the concept of 'tool-object reversal', as outlined by Virkkunen (2013, p.18). Here, the tool, which is the e-textbook in this context, inadvertently becomes the object (primary focus) of the system, overshadowing the original object, which is learning Physics. Instead of being a facilitative mediating artefact, the e-textbook becomes an entity to be learned and mastered in itself, diverting attention from the core academic objective.

The primary contradiction stands out as the most enduring and resistant among all the contradictions detailed in this chapter. It is central to understanding the dynamics of how and why students interact with e-textbooks, particularly when considering the concept of use value versus exchange value (Foot, 2014). The use value of an e-textbook is evident in the authentic epistemic growth of students, which contrasts with its exchange value, exemplified in students' adherence to institutional norms and policies.[1] When considering the use value, it is also worth weighing the utility of the tool against whether the students enjoy using it, as this influences their engagement and the nature of their learning experience.

Addressing the primary contradiction will require that the community offers support digital-literacy support if e-textbooks are used. It will also require that lecturers clarify the primary focus of the course, ensuring that there is an appropriate balance between technological proficiency and content mastery.

Limited Usage of e-Textbooks Versus Investing Energy to Explore the Functionalities of e-Textbooks

The secondary contradiction, symbolised by B in Fig. 4.1, raises tensions between the mediating artefact and subject of the activity. The mediating artefact is the e-textbook as embedded in the LMS. The subject of this activity is the group of first-year students taking the Physics course, who are registered in the Chemical Engineering and Maritime Studies Departments in the Faculty of Engineering and the Built Environment (FEBE) at the selected university of technology. The students' lack of knowledge about how to use e-textbooks leads to limited usage of them. The primary contradiction discussed above (where the object essentially becomes learning to use the tool instead of learning Physics), leads to this secondary contradiction, which is that the more time and effort students must invest in mastering the functionalities of e-textbooks, the less inclined they are to use these digital resources.

This secondary contradiction manifests as a dilemma centred on the students' struggles with the affordances and functionalities of e-textbooks, which hinders their learning process. For example, Student 3 confessed, 'I will be honest, I wasn't aware it had an index, and many things. I thought that maybe it's only for searching the chapters only, not any other searching things… Now, I know.' Similarly Student 4 described their lack of awareness of a specific feature in the following way:

> I don't think there's something wrong with the colour of hyperlinks or anything else, it's just that I was never informed that pressing there gives you additional information. So regardless of the colour green, blue, red, I would still leave it and continue reading the text. So, I just was not informed about it.

Another example of students having limited knowledge of the functionalities of e-textbooks is evident in an explanation given by Student 13:

> The reason I use my book [paper book] is that I have eye problems. So, when I look at the screen for a long time, I tend to have tears coming up. So, like I make notes, and then watch videos, then turn off the screen, get to my book.

---

[1] In CHAT, a primary contradiction could be a reflection of the fundamental tensions inherent in capitalist societies. According to Foot (2014), is located primarily within the object of the activity system, where is a direct tension in capitalist economics between use (for societal benefit) and exchange (for personal benefit), which give rise to a primary contradiction.

This shows that this student is unaware that issues such as eye fatigue from digital screens or physiological issues can be resolved or reduced by adjusting certain settings on the computer or e-textbook, such as reducing the screen brightness, setting a black background, changing the font and size of the text, and using the read-aloud function. Because of their limited knowledge of these functionalities, students default to a preference for what is more familiar.

This contradiction can be resolved as suggested by Student 7, 'If we have instructions on how to use an e-textbook, then it's acceptable because it's simple and straightforward', with which Student 2 agreed. Both these students contend that a lack of instruction on how to use e-textbooks was the main reason for not using e-textbooks. Similarly, Student 14 responded:

> [in the beginning], I … [didn't] know how to use a digital book. It's [in fact] much easier to use … if I want to highlight something, I just highlight it. … if I could … [be taught] more about it, a digital book … is much better.

Students mentioned that they did not have enough time and experience to study with e-textbooks. As Student 1 said, 'I would love to know how it's actually my first time using [e-textbook for study] so I don't know how to do so…'

The absence of clear and direct instruction on using e-textbooks for study purposes may imply a superficial and therefore simplistic use of digital-literacy practices when using e-textbooks, which may lead to the inappropriate use of e-textbooks, the perception that e-textbooks do not assist in acquiring knowledge, and, in some cases, a resistance to using e-textbooks.

This can be solved by providing more detailed and explicit instruction on how to use e-textbooks and facilitated practice and by embedding the e-textbook into the course in a better way. The more adept students become at using the e-textbooks as a tool, the more they will use them.

Contradiction C: Desire for Unlimited Access Versus the Restricted Distribution of e-Textbooks

The continuum of knowledge dissemination follows a fascinating historical trajectory, beginning with the transmission of wisdom through ancient oral traditions, which has endured across centuries. The subsequent emergence of script introduced a pivotal transition, enabling the recording and preservation of knowledge. This evolution reached a seminal juncture with the advent of the printing press, an innovation that catalysed an unprecedented proliferation of information dissemination. Now, since the creation of the internet, we navigate a digital landscape characterised by the swift interconnectedness of information. Free media on digital platforms has transformed the dynamics of how information is shared and accessed.

This last aspect holds particular importance in developing countries such as South Africa, where many first-year and first-generation students[2] face challenges

[2] A first-generation university student refers to an individual whose parent(s) did not obtain any tertiary-education qualification.

accessing educational materials and resources. Therefore a tension arises between the subject, namely the students, and the rules set by publishers regarding access (labelled as C in Fig. 4.1). This secondary contradiction manifests as a conflict, characterised by a disagreement between the desires of the students for unlimited access and the limitations imposed by copyright licences and restricted distribution that protect the interests of the publishers.

During an interview, Student 5 suggested, 'Access to the e-textbook. It must be an app like to read on the device. Can we download [publisher of the e-textbook] on its own and we don't have to go to the [LMS]?'. Student 1 added, 'Access to … e-textbooks available for everyone.' These statements by Students 5 and 1 indicate their preference for having unrestricted access to e-textbooks on their digital devices, as opposed to limited access under a publisher's licence and via the official system.

The restriction on knowledge distribution outside of the university reflects the traditional conceptions of ownership of intellectual property. Despite investments in open education resources (OER), resources for university students are limited in the subjects Science, Technology, Engineering and Mathematics (STEM) (Eaton et al., 2022).

Contradiction D: Preferences for Solitary Study Versus the Collaborative Study Affordances of e-Textbooks

While e-textbooks hold the potential to foster student collaboration, an underlying secondary contradiction (labelled as D in Fig. 4.1) surfaced between the subject, namely the students, and the mediating artefact, namely the e-textbook. This manifests as a conflict, stemming from the disparity between the collaborative learning affordances offered by e-textbooks and students' inclination towards solitary study.

Such preferences could stem from individual perceptions of the advantages of solitary study. It is also common for first-year students to be less accustomed to studying in groups for various reasons, such as individual preference, lack of prior experience with group assignments, or past dissatisfaction with teamwork.

During a focus-group activity, students were asked to read e-textbooks and engage in natural study behaviours, including studying with other students. Students 9 and 10 had similar preferences in that Student 9 stated that, 'I do not like studying in groups, so I'd rather study with myself', and Student 10 stated that he also preferred to 'study alone'.

However, despite Students 9 and 10 expressing a preference for solitary study, they engaged in collaborative activities that I, as the researcher, set up. They assisted peers upon request, which indicates a level of willingness to participate in shared learning experiences. While these students may not have naturally initiated such collaborative interactions, their actions still underline their capacity to contribute to collective learning and support their peers' educational journey.

Therefore students' collaborative actions within this controlled setup, despite their stated preferences, introduces a significant contradiction. The contradiction becomes more pronounced owing to the explicit opportunity for collaborative learning, which these students consciously chose not to fully embrace. This choice highlights a distinct tension between the collaborative potential embedded within e-textbooks and the students' inherent inclination towards independent study methods. The conflict is also evident in the students' potential lack of awareness of the benefits of collaborative learning, as well as their limited understanding of how e-textbooks can help them overcome the limitations that physical constraints have on collaboration.

In order to prepare the students to be able to collaborate in the workplace, collaboration is part of the curriculum and the students will have other opportunities to collaborate during their studies. However, currently the conflict between their preferences and the collaborative affordances of e-textbooks seems to fuel their resistance towards engaging in other collaborative learning opportunities with peers.

According to Engeström (1999), knowledge creation and learning can become collaborative endeavours when individuals begin to question accepted practices and concepts. This is carried out through individually and collectively undertaking epistemic action, such as questioning prevalent practices and ideas, analysing and modelling systemic causes of problems, modelling the new object and form of the activity, and putting the new model into practice (Engeström, 1999).

In light of this, students may begin to make use of the affordances offered by e-textbooks for collaborative learning if lecturers deliberately design activities that require shared annotation, co-constructed notes, or discussion around digital excerpts rather than treating the e-textbook as a static repository of content.

Perhaps it may be that students do not want to collaborate through the designated resource (e-textbook) because of the access and control issues discussed under Contradiction C, but that they happily chat with peers about curriculum content on unofficial apps which were not explored in this study.

Contradiction E: Pre-Curated Information in e-Textbooks Versus Unreliable Online Resources

When students started their studies at this university there was a notable shift in the mediating artefacts that they were expected to use. Previously the students had primarily relied on prescribed paper books but upon entering the university they were expected to transition to e-textbooks, embedded in the LMS, as the primary mediating artefacts (see contradiction E in Fig. 4.1). This tertiary contradiction manifested as a conflict because students often made this change reluctantly.

Tertiary contradictions are historically rooted (Engeström & Sannino, 2011), and such contradictions emerge within an activity system when a more 'culturally advanced' activity becomes the focus (Engeström, 1987:78). The introduction of a new mediating artefact can also herald the onset of a new developmental phase, wherein the activity system undergoes redefinition and reconfiguration. In the context of this study, this redefining moment was propelled by the students having to adopt new digital-literacy practices rather than simply relying on familiar and popular internet platforms.

Students explained that they relied on Google, YouTube and similar online platforms to clarify concepts. Student 1 said, 'So I'll use Google. YouTube is working for me. They explain the concepts…'. Student 4 said, '… Google explains the word and then sometimes Google gives you a play button whereby it says the word for you, and then that's how I know how to say [it]'.

However, while diverse online resources can offer students clearer insights and a richer understanding of concepts, not all information obtained online is trustworthy, and some students did not seem to realise this. For example, Student 7 said, 'It's well-known that Google gives accurate answers'. This is a cause for concern. In an age rife with fake news and an overwhelming influx of information, a pivotal literacy practice is the ability to discern, curate and fact-check to identify the most pertinent and trustworthy information.

Some students were surprised to learn that the e-textbooks give links to sites that give accurate and reliable extra information, or that they could search for definitions. Student 3 said, 'I was not aware that there was an option of just searching by definition, okay. So, it's a helpful function when you do not have enough time. That's great.'

Yet while the e-textbook curates reliable information, it also decreases the need for students to develop their own research curation skills. Learning how to curate research is critical for students' future success, both academically and professionally, and over-reliance on e-textbooks for pre-curated information may hinder students' opportunities to develop these skills. For example, from Student 2perspective: 'I watch videos that are random. Yeah, but not always because they sometimes confuse me. I get confused' and Lecturer 3 said, 'It's not useful to have them [different resources] To have many of them. Because for reference there are many ways to reference'.

Therefore, e-textbooks should be viewed as complementary resources rather than substitutes for developing research curation skills. As well as helping students use e-textbooks, lecturers must also incorporate research-curation assignments into the curriculum, guide students in the selection and evaluation of resources, and promote critical thinking and analysis. Additionally as suggestion, reflecting on the links that are used in the e-textbooks could be a way of starting a discussion about what is considered a reliable source. By doing so, students can benefit from the convenience of e-textbooks while still developing the skills necessary to succeed in their academic and professional lives.

Conclusion

This chapter set out to explain how e-textbooks can disrupt students' learning by shifting the role of the mediating artefact from a tool that supports activity to an emerging object in its own right. When the mediating artefact becomes the new object, students' attention can be drawn away from the curricular purpose of reading and understanding content and toward the work of managing, mastering or circumventing the digital tool. In this case, the primary contradiction between the disruptive novelty of the e-textbook interface and the students' existing reading practices and expectations functions as a foundational tension that influences all the other contradictions discussed. Because this contradiction is internal to the tool itself, between its potential to support meaningful learning and its capacity to generate frustration, cognitive load and anxiety, it creates cascading tensions across the other nodes of the activity system. As the analysis has shown, this primary contradiction often gives rise to secondary contradictions between tools and rules (where institutional imperatives to adopt e-textbooks collide with classroom practices), between tools and subjects (where students struggle to engage meaningfully through the interface), and between tools and division of labour (where lecturers now need to become technical facilitators). Tertiary contradictions also emerge when more culturally advanced digital pedagogies are introduced without adequate support or alignment, further disrupting the system. Because the nodes are interconnected, these contradictions and the resulting tensions have implications that extend beyond the individual nodes and affect the entire activity system.

The discursive manifestations of contradictions framework was useful in making these dynamics analytically visible. Rather than treating contradictions as abstract theoretical categories, the framework helped trace how contradictions surfaced in students' talk, in their coping strategies, and in their affective responses. In this regard, discursive manifestations provided an empirical route into a phenomenon that is often described conceptually in CHAT literature. It also allowed the analysis to proceed from language to system, moving from what students say, do and feel to how the activity system is organised and strained. It is true that this chapter focused more on the levels of contradiction (primary, secondary, tertiary) than on the typology of contradictions outlined by Engeström, and this is an area that could be elaborated further. Nonetheless, identifying levels was methodologically useful for demonstrating how the primary contradiction propagated throughout the system, thereby illustrating the systemic reach of disruption.

It is important to note that disruption is not solely negative. E-textbooks are often framed as improvements over printed texts because they allow readers to search, annotate, hyperlink, highlight, and, crucially, to collaborate and share information efficiently. These affordances can support new modes of engagement and create opportunities for more interactive, networked, and participatory forms of learning. However, the very same artefact can constrain behaviour by imposing new expectations, requiring new competencies, and generating new forms of anxiety. Students often inhabit this tension simultaneously: they appreciate the increased convenience and collaborative potential of e-textbooks while also feeling disrupted by the technical, cognitive and affective demands they introduce. This combination underscores the central idea captured in the title: when a mediating artefact becomes the new object, the activity does not simply continue with a better tool; it is reconfigured in ways that are ambivalent, multi-faceted and deeply systemic. The contradictions documented here are therefore not only evidence of disruption but also indicators of developmental possibility, suggesting that future iterations of digital tools, pedagogies and institutional practices could transform disruption into genuinely expansive learning.

References

Eaton, D. C., Bonner, K., Cangialosi, K., Dewsbury, B., Diamond-Stanic, M., Douma, J., et al. (2022). Sustainability and justice: Challenges and opportunities for an open STEM education. *CBE—Life Sciences Education, 21*(3), es4.

Engeström, Y. (1987). *Learning by expanding: An activity-theoretical approach to developmental research*. Orienta-Konsultit.

Engeström, Y. (1999). Innovative learning in work teams: Analyzing cycles of knowledge creation in practice. In Y. Engestrom, R. Miettinen, & R.-L. Punamaki (Eds.), *Perspectives on activity theory*. Cambridge University Press.

Engeström, Y., & Sannino, A. (2011). Discursive manifestations of contradictions in organizational change efforts: A methodological framework. *Journal of Organizational Change Management, 24*(3), 368–387.

Foot, K. A. (2014). Cultural-historical activity theory: Exploring a theory to inform practice and research. *Journal of Human Behavior in the Social Environment, 24*(3), 329–347.

Virkkunen, J. (2013). *The change laboratory: A tool for collaborative development of work and education*. Springer Science & Business Media.

References

Chapter 5
The 'Culturally More Adept Others' Who Prescribe e-Textbooks: How Are Lecturers at a University of Technology Navigating Their Evolving Role?

Abstract This chapter investigates how lecturers at a university of technology navigate their evolving pedagogical role when required to prescribe e-textbooks as primary learning resources. Using Cultural-Historical Activity Theory (CHAT) as an analytical lens, the chapter examines how contradictions manifest within the lecturers' activity system and between intersecting systems when digital texts become institutional expectations. Empirical findings demonstrate that e-textbooks reshape lecturers' identities from content experts to digitally mediated facilitators, generating tensions around responsibility, authority and pedagogical autonomy. Additional contradictions emerge from the disjuncture between foreign-produced e-textbooks and local curricular, linguistic and socio-cultural contexts, raising questions about epistemic relevance and inclusivity. Device limitations, infrastructure constraints and uneven digital literacy further influence the enabling potential of e-textbooks. Beyond the lecturers' system, contradictions also arise between institutional procurement processes, commercial publishers and classroom realities, illustrating the relational complexity of digital adoption. The chapter then considers the extent to which e-textbooks assist lecturers in meeting disciplinary demands in Physics, particularly regarding conceptual clarity, scaffolding, and visualisation of abstract phenomena. A final concern relates to whether students' engagement with e-textbooks adequately bridges theory gaps. The chapter concludes that lecturers occupy a critical mediating position in digital transitions, requiring support, contextualisation and agency to fulfil their evolving role effectively.

Introduction

When lecturers prescribe e-textbooks for their students this affects the role of the lecturer in many ways, both freeing lecturers from certain responsibilities and encumbering them with new ones. The resulting contradictions in the lecturers' activity system, and in the way that this intersects with the students' activity system, indicate that the lecturers are experiencing changes to their role. They have preferences about how the division of labour will shift, but these preferences do not always

J. Hardman, E. Rzyankina, *Contradictions in Cultural-Historical Activity Theory*, https://doi.org/10.1007/978-3-032-28650-5_5

align with institutional expectations, technological affordances or students' needs. As a consequence, lecturers must negotiate new boundaries around what they consider to be core pedagogical work, what can be delegated to the digital tool, and what emerges as unintended administrative or technical work. These negotiations reveal that the adoption of e-textbooks is not merely a change in educational technology, but a reconfiguration of the activity system in which lecturers operate, with implications for identity, agency and pedagogical intent.

Although beyond the scope of this chapter, one could even weave in a mention of Socrates, whose questioning style represents an early and influential form of inquiry into assumptions and taken-for-granted meanings. However, whereas Socratic questioning tends to narrow the lens by exposing contradictions in individual arguments and thereby clarifying definitions, the kinds of questions asked in this chapter tend to widen the lens by situating contradictions within broader social, pedagogical and technological activity systems. In this respect, the Socratic method illuminates internal logical tensions, while a CHAT-informed analysis foregrounds systemic tensions that arise when new artefacts, roles and expectations reorganise collective activity. The comparison is useful because it underscores that not all contradictions are of the same order: some are conceptual, elicited through dialogue, whereas others are structural, generated through the reconfiguration of tools, rules and divisions of labour.

How Do Contradictions Manifest in the Lecturers' Activity System When e-Textbooks Are Used?

In this chapter, I discuss the interviews conducted with four lecturers and present various contradictions that emerged. All the contradictions I analyse in the lecturers' activity system are secondary[1] or tertiary contradictions,[2] and are labelled A–C in Fig. 5.1.

These contradictions could lead to improvements in the activity (see Chap. 3), but not all potential improvements will be implemented and, even if they are, may take time to become established in the system. In the sections below I examine the complexities that e-textbooks introduce in academia.

[1] Secondary contradictions occur between nodes of the activity system, for example between 'community' and 'mediating artefact'.

[2] Tertiary contradictions occur between an earlier version of an activity system and the current version of that activity system, for example before and after implementing a new mediating artefact such as an e-textbook.

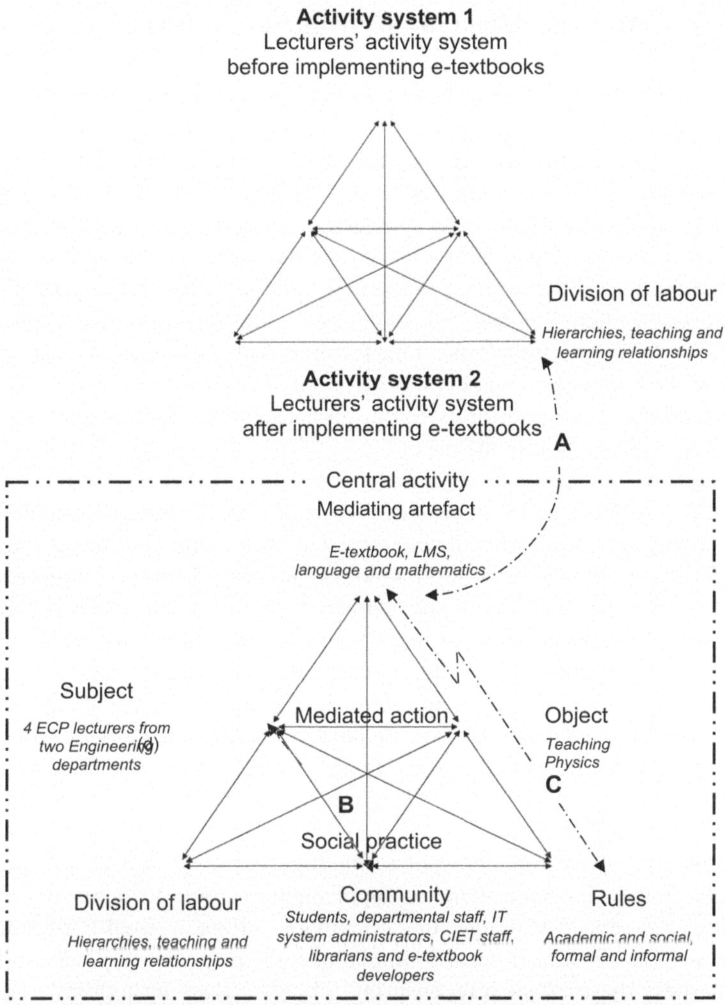

Contradiction (presented in the form of a question)	Type and location of contradiction
A: How do e-textbooks affect the role of the lecturer?	Tertiary contradiction between activity system 1's division of labour and activity system 2's mediating artefact
B: Are the foreign e-textbooks suitable for the local context?	Secondary contradiction between the subject and the community (in activity system 2)
C: Are e-textbooks fit for purpose in the face of device disparities?	Secondary contradiction between the object and the rules (in activity system 2)

Fig. 5.1 Three contradictions in the lecturers' activity system

How Do e-Textbooks Affect the Role of the Lecturer?

The paper textbook was the material tool in the mediating artefact in the previous version of the lecturers' activity system, and it has been replaced in the new version of this system by the e-textbook. However, the division of labour in the new version of the activity system has not adjusted to accommodate the new mediating artefact. This leads to a tension that can be considered a tertiary contradiction, because it exists between the division of labour in the previous version of the lecturer's activity system and the mediating artefact in the new version of this activity system. This contradiction, labelled A in Fig. 5.1, underscores the transformative impact of the e-textbook on the conventional role of the lecturer. With the advent of the e-textbook, lecturers are no longer just content presenters—if, indeed, they ever were. They now face the challenge and opportunity to design and curate enriched learning experiences—even more so than in the past.

Lecturer 2 notes the support that the e-textbook offers to lecturers, saying, 'E-textbook makes life so much easier in terms of understanding the content yourself, then being able to deliver the content, and then being able to pursue assessment. It's brilliant.' However, while e-textbooks can bolster lecturers' comprehension, they still bear the responsibility of conveying content to students in a comprehensive manner using the semiotic tools of language and mathematics, which are also part of the mediating artefact. The lecturer must intertwine the content with context, insights and structure. Effective teaching is a blend of pedagogical strategies, instructional design, student engagement and clear communication. For this reason, lecturers will remain important as the e-textbook facilitates the development of the traditional lecturing model into a culturally more advanced, student-centric approach.

The objective of the university of technology is not to replace lecturers with e-textbooks, but to leverage them as supplementary tools to demystify intricate engineering concepts. Although the e-textbook offers a wealth of additional resources such as visualisations and simulations, the responsibility falls on the lecturer to harness these effectively, ensuring a holistic learning journey. Lecturer 3 emphasises this need for balance:

> I think e-textbooks can definitely assist you in your teaching ... It's there to support your teaching, not take over. Those lecturers who just use the e-textbook to teach on their behalf need to be cautious.

This highlights a risk: while e-textbooks offer undeniable advantages, relying solely on them could erode the lecturers' active involvement in teaching which could lead to a drop in the quality of the learning experience. E-textbooks could thus lead to diminishing lecturer–student interaction. Lecturer 3's comment illustrates the importance of lecturers' maintaining agency and control over the instructional process.

A tension is apparent between the capabilities of e-textbooks as intermediaries and the conventional roles of lecturers. On the one hand, e-textbooks offer tools and resources that can simplify the teaching process, as articulated by Lecturer 2. On the other hand, there is a realisation, particularly from Lecturer 3, that relying solely on e-textbooks can diminish the active role of a lecturer. This manifests as a conflict.

Lecturer 1 raises another dimension of the contradiction by stating that using the e-textbook platform 'solely for assessment overlooks its potential for enhancing student learning.' This indicates that e-textbooks should not be treated only as assessment repositories; they should act as dynamic platforms that aid student performance, and facilitate and monitor various levels of lecturer–student engagement. If lecturers use e-textbooks only for assessment, they miss out on their pedagogical benefits, such as their support for scaffolded learning. However, if they lean too heavily on e-textbooks, they risk diminishing their active role and possibly the quality of student–lecturer interaction. This indicates a double bind, where lecturers seem trapped between two conflicting constraints.

Historically, the role of a lecturer has been understood in various ways—from a 'sage-on-the-stage' to a learning facilitator. The e-textbook has the potential to shift this role towards the latter, and there is a trend of students using e-textbooks for independent study. Scholars such as Ahn (2014) and Gu et al. (2015) have even argued that e-textbooks can foster self-regulated learning. However, self-regulated learning refers to the process where students personally initiate and manage their own learning processes. It is not just about the ability of students to study independently, but rather their ability to understand and control their own learning. Therefore, while e-textbooks may support self-regulated learning by offering flexible and interactive affordances, they do not in themselves produce self-regulated learners. The motivation, metacognitive capacity and learning orientation of the student remain decisive, and a motivated student could regulate their learning using traditional resources just as an unmotivated student could fail to regulate their learning even with advanced digital tools.

There is anxiety that this supplementary tool could undermine lecturers' authority When asked whether the e-textbook could replace the lecturer, Lecturer 2's initial response was to say, 'Obviously, I'm a teacher, I never want to agree with it that something can replace me, but the real answer is "yes".' But further discussion led to this lecturer clarifying, 'So, I wouldn't say replaced, because I feel like some things can be supplemented.' This seems to be a contradiction manifesting as a dilemma, because there are the 'Expression of incompatible evaluations'.

Lecturer 2 elaborated on this by saying:

> there are some students who will never be able to function in a classroom, but if you give them a laptop, you give them access to everything. And they still want to be in the classroom being taught something.

This suggests that the situation is student dependent, with some learners thriving in a digital environment while others want a lecturer present to lead the learning process.

Lecturer 4 echoes this theme when describing his role as follows:

> lecturer comes in with more effort, especially to cover the background [engineering problem] for the students on what they've been exposed to help them use their imagination. So, e-textbook is linked to content and doesn't really stretch the minds of the students to a point

> where if you want to make an example about a particular concept, you want to make an example and you want to bring a typical application of that particular concept. So, an e-textbook won't do that for you.

This shows that e-textbooks, while they are rich in content, do not currently compete with the lecturers in how they contextualise concepts and apply them in relevant ways. From a Vygotskian perspective, this gap can be understood through the Zone of Proximal Development (ZPD). The ZPD is not only about assisting learners to accomplish tasks that are marginally beyond their independent capability; it also involves creating the conceptual bridges that allow students to make lateral connections across ideas so that learning becomes meaningful rather than merely accumulative. Lecturers routinely operate within this space by unpacking concepts, drawing cross-disciplinary parallels, offering examples and counterexamples, and connecting abstract ideas to concrete experience. These forms of mediation activate the ZPD by enabling students to reorganise knowledge and integrate it into a broader cognitive framework. E-textbooks, by contrast, tend to present content as vertically structured and internally coherent, but they rarely scaffold the lateral, relational work that underpins conceptual development. As a result, even motivated and capable students may struggle to achieve the kind of conceptual movement that typically occurs through guided participation within the ZPD.

A notable observation shared by multiple lecturers is the evolving nature of their role. This evolution raises two possibilities: firstly, the potential diminishment of the lecturer's unique role, and, secondly, the prospective enhancement in teaching quality when e-textbooks are optimally utilised. From a CHAT perspective, it is crucial to consider the cultural and historical aspects of teaching. Some lecturers view the incorporation of e-textbooks as a progressive step, leveraging technology to complement and enrich their instruction (Pepin et al., 2015). Conversely, certain lecturers perceive e-textbooks as a constraint, feeling a reduction in their agency and autonomy. This sentiment might stem from the sense of control they possess in traditional teaching contexts.

Lecturer 3 shows that a 'culturally more adept other', to use Vygostsky's term, is still needed so that the activity system functions:

> Just because there's an e-textbook doesn't mean you can stop teaching, lecturing, and explaining what is happening within that topic. Otherwise, there is a disconnect between the students, the lecturer, and the e-textbook. So, you have to make sure that everybody's working together, and the e-textbook contributes to your teaching, but it doesn't take over your teaching.

Furthermore, Lecturer 2's words indicate that students feel that they deserve a 'culturally adept other' who is actively involved:

> I found that even with my final year students, when you get to chapters that maybe have a lot of theory… And when you direct them to a resource, even if it's a video or something else, the perception that they feel like you're not teaching, you're trying to get away with teaching and you're being lazy, or it's not what they pay for, or those types of vibes, or reactions, or things like that.

It seems that students' expectations are rooted in traditional methodologies, and that they do not want the mediating artefact to disrupt the established activity system in a way that makes them have to learn without the lecturer's mediation. This may also be because, during COVID-19, students had been told to do tasks without adequate guidance on how to manage their own learning effectively.

Teaching and learning under social distancing conditions, brought about by the global pandemic from 2020 onwards, highlighted how technology could be used in remote education and increased the interest in e-textbooks. Lecturer 3 captures the essence of the prevailing sentiment:

> You are kind of influenced by what is happening around you – for example, the fact that e-textbooks are becoming so popular and common. The pandemic where we couldn't teach face to face. So, it definitely changes the way you teach and your teaching philosophy because it makes you think about how to make your teaching methods better.

Therefore the popularity and ubiquity of e-textbooks were part of what made Lecturer 3 reflect on how she could improve her approach. In this way the e-textbook is triggering transformation, prompting lecturers to design more meaningful learning experiences.

Clearly contradiction A between the mediating artefact (e-textbook) and the division of labour, is multi-layered, marked by a conflict concerning the e-textbook's potential, a double bind concerning how much lecturers should utilise it, and a dilemma for the lecturers about whether it could replace them. Recognising and addressing these tensions can foster a productive integration of e-textbooks within the educational landscape. Resolving them requires a balanced approach and professional development, where training programmes equip lecturers with the necessary skills and strategies to effectively design and facilitate learning experiences that optimise the use of e-textbooks.

How Suitable Are Foreign e-Textbooks for the Local Context?

Contradiction B in Fig. 5.1 exists between the subject and the community (referring life experiences of the students) within the lecturers' activity system. This secondary contradiction manifests as a conflict stemming from the contrasting perspectives and the criticisms voiced by the lecturers. The core of the disagreement revolves around the contextualising of content in the e-textbooks. Historically, many textbooks, whether paper or digital, have been centred around American and European contexts, both physical and cultural. As a result, students from other backgrounds, particularly from the global South countries[3] such as South Africa, often find it challenging to relate to the content. For instance, Lecturer 2 points out:

---

[3] The Global South countries are countries that are characterised as having developing economies. They are primarily located in the southern hemisphere.

> Some of the examples in e-textbooks are American context examples. You need to come up with your own South African [examples], or ones that will be easily understood by people who are in our context. But it's a lifesaver because compared to subjects where you don't have an e-textbook, you have to go look for examples in the textbook.

Lecturer 1 further elaborates on the cultural disconnect:

> The context in the e-textbook is written mainly for the European side, or they're from America. This might confuse a first-year student. For example, snow and sliding on the snow. I'm in South Africa … we've never had snow … you've never seen it … if the context of an example is created in such a way that I relate to the context, I formulate meaning from my own language. To translate to English is not an issue, but to start it from there and bring it back to me. And if I cannot relate to those kinds of things, like snow. This will make it very difficult for [a student] to get these meanings.

Such examples underscore the inherent tension within the lecturing activity system. Students are not only required to understand the technical content; they must also understand unfamiliar cultural references.

This cultural misalignment brings up the broader issue of decolonising content. There is a strong push to make educational content more contextually relevant and inclusive. Lecturer 1 and Lecturer 2, although acknowledging the benefits of e-textbooks, emphasise the need for content that resonates with students' lived experiences (see Chap. 2 for concept *perezhivanie and/or dramatic event).* Their criticisms highlight the challenges e-textbooks present, particularly in their lack of relevance to the local context.

In the Global South, the mediating role of lecturers extends beyond disciplinary expertise and instructional design; lecturers frequently act as cultural translators who help students navigate not only the conceptual terrain of a subject but also the cultural logics embedded within knowledge systems, assessment practices and institutional expectations. This requires lecturers to be culturally adept within their own contexts, attuned to linguistic diversity, socio-economic disparities, and locally situated ways of knowing, while also interpreting and contextualising curricula that are often imported from Global North epistemic traditions. In this dual role, lecturers bridge cultural and pedagogical distances by providing examples, metaphors and problem framings that resonate with students' lived experience, thereby enabling learning to take root within culturally meaningful frames of reference.

However, the issue is not just about content relevance but also touches on language. While Lecturer 1 does not complain about the e-textbook being in English, they do say that 'using an e-textbook, if learners are encouraged to formulate meaning from their own language first, that could really work.' This shows that lecturers can play an important role in encouraging students to grasp a new concept by having those students articulate it in their home language.

While e-textbooks introduce new possibilities for teaching and learning, their current implementation falls short of addressing the embedded contradictions in the lecturing activity system. The challenges highlighted by the lecturers provide valuable insights. By integrating diverse examples and prompting students to derive meaning from their own language and experiences, e-textbooks could support a more profound connection between students and content. Currently the lecturers do this as part of the role they play. This connects directly to Contradiction A, because while e-textbooks are introduced with the assumption that they can replace or reduce the lecturer's mediating role, the reality is that they do not perform this cultural translation. Addressing these challenges can harness e-textbooks' potential to alleviate tensions in the system, promising a more inclusive and effective learning experience for all. Perhaps publishers could address these challenges if they incorporated direct feedback from students and lecturers into the future development of e-textbooks.

How Enabling Are e-Textbooks in the Face of Device Limitations?

Contradiction C in Fig. 5.1 is a tension between the mediating artefact, which is the e-textbook, and the rules of the lecturers' activity system, which try to ensure equal access to resources in a very unequal socio-economic context. This secondary contradiction manifests as a conflict, which became particularly evident during the lockdown period of the pandemic, when students were no longer on campus, and teaching and learning transitioned to being an online activity. This is because remote education revealed discrepancies in the students' readiness to adopt e-textbooks.

Lecturer 2 highlighted the different socio-economic levels of students, and the effects of these:

> There's different levels of marginalisation... if you're looking at the level where some students they [either] come from a small town or a good public school or a moderate public school. And they get here, they know how to do basic things, like search on, they've got a smartphone, so they can kind of figure out from a smartphone to a computer, or they can learn it quickly, then access to an e-textbook is wonderful.
>
> ... there's a degree of marginalised students that ... have never seen a computer, [they have] some smartphones that are considered smartphones, but don't have a PDF [app], or don't have space to download. So then giving them access to an e-textbook is not going to do much for them because they still can't access it at all.
>
> ... So in those cases, then making an e-textbook as a resource that they need to use, makes then the gap bigger...

In essence, students who did not possess the necessary technology were unable to access the e-textbook, and this was exacerbated by the digital divide in their backgrounds. This problem was widespread because, as Lecturer 1 explained, 'at [the university of technology] we teach poor students'.

Lecturer 1 discussed the problems that surfaced when e-textbooks were initially introduced to first-year ECP students. Lecturer 1 had originally believed that subsidising e-textbook costs would resolve accessibility issues, especially for students living off-campus who face challenges with internet access:

> My motivation was to ensure that I was harmonising my students so that they were all on the same level. So, to be able to pay for that platform, and it is the e-textbook insight that they can read, that at least everyone could have access to the e-textbook.

However, by resolving the contradiction of equity of access to e-textbooks, a new contradiction became evident, which was that many students lacked internet access as well as an appropriate digital device, owing to deeper socio-economic challenges. Lecturer 1 explained:

> My thought was that if I cut the cost, but paid for them, I'd solved the problem. But, in the end, I realised that the impact of data if I had done that and give the student data to those who were staying off campus and were able to enter a small amount of data could have been different, but the fact that there were no laptops was discovered.

Lecturer 4 echoed these challenges:

> While many students staying in townships, they did not have internet access… They could say, 'Sir, we don't have the internet at home, and we don't have this and that we can't use it [e-textbook]'.

Merely providing e-textbooks does not resolve all the underlying issues of technological and socio-economic disparities for South African students: access to the internet and data, as well as to digital devices such as smartphones and laptops, pose significant challenges for many students.[4] Lecturer 1 summed up the problem of e-textbooks by saying, 'not having access to digital devices means not having access to knowledge'.

Lecturer 1 intervened on the students' behalf:

> I tried to address this when I went to the head of ECP at the time and asked if we could give students in the ECP laptops, and I got a lot of resistance; he said: 'We don't want an ECP to be a place where people come in and get what they want.'

Nevertheless, during the pandemic the university of technology provided laptops as Lecturer 1 had asked:

> When COVID-19 came through, I looked around and saw that everyone had started giving them computers. I said I'd seen this and I have great empathy. I saw the future before it arrived. We will, as I said, arrive, because I was told that resources existed but people refused to use them.

Lecturer 1 noted that once necessary tools such as laptops and data were provided there was enthusiasm for the e-textbook platform among students:

> I mean, [during] COVID time [moving teaching from the traditional classroom to online teaching and learning] was the smoothest thing [because] … they [students] got laptops and data. So, the complaints that I heard at the time were that not everyone has it, so why can't we use [name of the publisher platform of e-textbook] for other subjects?

Lecturer 4 agreed:

> Students were happy to practise with it because they had the institution's data and laptops. So, there were no complaints … Since they [students] have been equipped with all the necessary tools like data, laptops, I did not have any complaints.

---

[4] We also need to note that students need time to become familiar with how to use an e-textbook. This requires that the lecturers should be ready to intervene and assist in mediating access to these textbooks.

Lecturer 2 acknowledged the ownership of cell phones and laptops had become widespread among students, even if they did not have internet access at home, and indicated that the digital device was the gamechanger in terms of making e-textbooks useful:

> Some of my students have to go to an internet cafe, but they have access to a laptop or computer to be able to access the e-textbooks, whereas if they didn't have access to e-textbooks, they wouldn't have access to that information.

So, even though the e-textbook was only accessible within the LMS, it had become usable. As Lecturer 1 noted, 'it will never be downloadable' but 'you can read the text'. Lecturer 4 observed that now e-textbooks provide valuable information to students that would be less easy to obtain through 'the hard copies located in the library'.

Therefore, the conflict was resolved by the university of technology providing equitable access to e-textbooks and digital learning resources. The tension that had existed between the mediating artefact and the rules regarding accessibility underscores the need for comprehensive strategies to bridge the digital divide and ensure equal opportunities for all students.

This dynamic can also be understood in terms of Leontiev's distinction between collective activity and individual action. In Leontiev's formulation, activity is oriented toward a shared object and is realised through a constellation of individual actions that may not, on their own, grasp the full motive of the collective endeavour. For the object of teaching Physics with e-textbooks to be achieved, the activity cannot consist solely of the isolated action of supplying digital texts. Someone must be aware of how the actions of different actors, publishers supplying content, IT units managing platforms, lecturers designing pedagogy, and administrators setting policies, will coalesce into an enabling system. The action of supplying e-textbooks therefore needs to be accompanied by the action of supplying devices, data and support; otherwise, the collective activity cannot realise its object. Framed this way, the lecturer's advocacy for devices is not merely an individual preference but an attempt to align dispersed actions so that the collective activity system can function coherently.

How Do Contradictions Manifest Between Intersecting Systems When e-Textbooks Are Used?

Thus far this chapter has provided a discussion of secondary and tertiary contradictions that manifested in the lecturers' activity system. In this section, I use third-generation CHAT (see Chap. 1) to discuss quaternary contradictions present between the students' activity system and the lecturers' activity system. Third-generation CHAT uses joint activity as the unit of analysis rather than individual activity (Daniels et al., 2010), and the focus is on interaction between two activity systems through a jointly negotiated object.

To achieve this jointly negotiated object, third-generation CHAT includes subjects and community members from both activity systems, who will have diverse traditions, interests and points of view. It assists researchers to see how participants in a study construct the object of their activity differently. It also foregrounds common contradictions, which give rise to tensions between these activity systems, in this case in relation to how the e-textbook is utilised.

In Fig. 5.2, activity system 1 is the students' activity system and activity system 2 is the lecturers' activity system. Figure 5.2 demonstrates the connections between the two activity systems and identifies the object of each of them. For the students, the object is learning Physics and for lecturers the object is teaching Physics. Object 2 is a potentially shared object that is constructed by both activity systems; it is teaching and learning Physics in Engineering.

The primary contradiction (see Chap. 4, Fig. 4.1, Contradiction A) within the node of the object in the students' activity system will also affect the shared Object 2 of the intersecting systems (see Fig. 5.2). This is because Shared Object 2 emerges through the interaction of the two systems and therefore carries forward the tensions embedded in each of them. When students experience uncertainty, anxiety or misalignment between the expected and actual use of the e-textbook, this tension propagates into the shared space of teaching and learning, shaping how

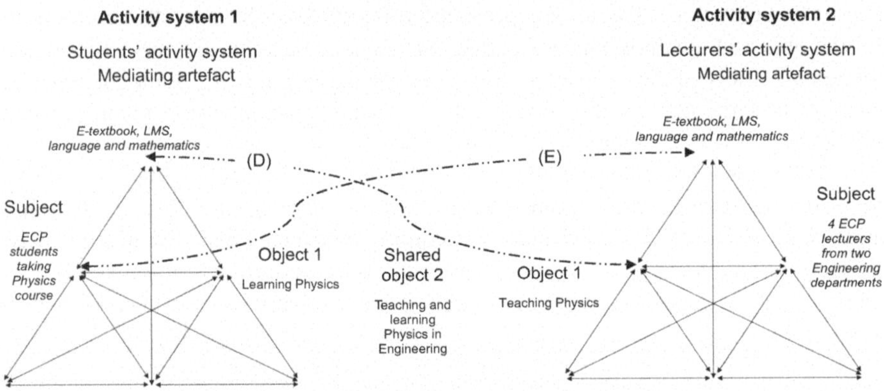

Contradiction (presented in the form of a question)	Type and location of contradiction
D: How will the e-textbook be useful to lecturers in helping them to meet their responsibility of teaching Physics?	Quaternary contradiction between the mediating artefact in activity system 1 and the object in activity system 2
E: How well do es reading the e-textbook fill students ' theory gap?	Quaternary contradiction between the subject of activity system 1 and the mediating artefact of activity system 2

Fig. 5.2 Two contradictions in the intersecting activity systems of students and lecturers

lecturers must respond and how the joint activity unfolds. However, once both students and lecturers become proficient in using the e-textbook and fully understand its affordances, the way that they attempt to achieve the shared object may shift. Proficiency may reduce the disruptive elements of the primary contradiction, enabling a reconfiguration of roles, tools and pedagogical strategies in pursuit of the shared object. In such cases, the shared object may become less about coping with the tool and more about leveraging the tool for deeper disciplinary engagement, potentially opening space for expansive learning rather than merely compensatory adaptation.

How Will the e-Textbook Be Useful to Lecturers in Helping Them to Meet Their Responsibility of Teaching Physics?

The primary contradiction in an activity system exists within a node and forms the basis for the other levels of contradiction. The primary contradiction in this study (discussed in Chap. 4) is learning Physics with e-textbooks versus learning how to operate the e-textbooks, and it exists within the object of the students' activity system, which is learning Physics. This primary contradiction influences the quaternary contradiction that involves how the e-textbook will be useful to lecturers in helping them to meet their responsibility of teaching Physics.

This quaternary contradiction exists between the activity systems of the students and lecturers. In Fig. 5.2 a line labelled D signifying this contradiction connects these two systems. The tension arises between the mediating artefact, namely the e-textbook, in the students' activity system and object 1 of the lecturers' activity system, namely teaching Physics. This tension exists because in order for the e-textbook to help lecturers discharge their duties to students, someone has to instruct the students on how to use e-textbooks. This manifests as a conflict between the students' need for instruction on e-textbook usage and the lecturers' belief that it is not their responsibility to provide such instruction.

Lecturers believe that publishers should offer videos or tutorials to guide students so that they could navigate the e-textbook and leverage its resources effectively. The incorporation of explanatory videos detailing the e-textbook's features would serve as instructional tools, empowering students to understand and decide on the utility of these features. Lecturer 2 suggested:

> if I were to add something, I would add that feature, maybe some videos explaining the different features they have, so the student can go on the video, see what it is, know how to do it, and then decide whether or not to use it.

While e-textbooks offer various advantageous features, there is a concern regarding the lack of instructions on how to utilise specific functions within the platform. Lecturer 1 observed that, without proper explanation and guidance, students may not fully benefit from e-textbooks and that they 'might end up just reading the textbook in digital format'.

While lecturers feel it is the publishers' duty to offer this guidance, students expect guidance from their lecturers. Lecturer 3 commented:

> They [students] are scared of trying something new. But as soon as they do, it's like, "Oh, this is great. Why haven't I been using it all the time?" So, I think they need an introductory class.

Lecturer 3 did not place the responsibility of teaching students to use e-textbooks only on the publishers, and recognised that lecturers had a part of play in this, but only suggested that a single lesson be given over to achieving this. More time is probably needed from lecturers, or other university staff members, to assist students in transitioning from conventional print reading to digital reading practices. The effectiveness of e-textbooks hinges on a community that supports their use, and the lecturers are the paid professionals that have the responsibility to see that adequate instruction is provided to students.

How Well Does Reading the e-Textbook Fill Students' Theory Gap?

The other quaternary contradiction, labelled as E in Fig. 5.2, is the tension that arises between the subject of the students' activity system and the mediating artefact of the lecturers' activity system. The core issue is that e-textbooks are intended to emphasise comprehensive engagement, including reading, while Engineering students exhibit behaviour that prioritises calculations over reading. This discrepancy between the educational intention and student behaviour forms the basis of a conflict.

Engineering students demonstrate a marked preference for calculations over in-depth reading. This might not just be confined to the realm of e-textbooks, but could be part of a broader trend. It could also hint at a preference for tangible, everyday practical knowledge over abstract, theoretical knowledge. Lecturer 2 describes the academic inclinations of the students:

> Engineering students generally don't like to read … Generally, they skip over the part that needs reading. And they go straight into the calculations. I don't know why. I think that could be something that can really hinder the implementation of e-textbooks if used inappropriately.

Engineering students typically lean towards solving 'real-world' problems and design solutions, which tend to involve calculations, simulations and hands-on work. This focus on application might lead students to prioritise direct engagement with calculations and problem-solving exercises, which they see as more directly relevant to their future roles as engineers. Consequently, they might perceive reading, particularly theoretical or scientific content, as secondary or less immediately beneficial. This behaviour could reflect a perceived hierarchy of knowledge where applied knowledge is prioritised over theoretical understanding. This distinction might extend beyond e-textbooks to other forms of educational materials as well.

In addition, evidence suggests that students in general often lean towards multimedia content, such as videos, over traditional textual content in e-textbooks, once they know how to navigate an e-textbook. Student 14 explained, 'I chose to watch the first video, yeah. I think videos for me; it is mostly listening to videos.' This shows that this student starts engaging with an e-textbook through watching videos (theory and practical examples of calculations), and that their use of its offerings does not extend far beyond this. However, student behaviour can vary based on individual learning styles, teaching approaches and contextual factors.

Lecturers need to find ways to encourage students to engage with all the e-textbook's resources that will help them learn better. This may entail directing students to watch videos on theoretical content, as both theorical knowledge and practical application are important aspects in Engineering education.

Conclusion

E-textbooks usher in opportunities for pedagogical innovation, but they also present new challenges. The primary contradiction in the students' activity system consists of two aspects that are in opposition: learning Physics with e-textbooks versus learning how to operate the e-textbook. The effects of this contradiction ripple through the way the students' activity system intersects with the lecturers' activity system, as well as through the lecturers' activity system itself. The primary contradiction: presents lecturers with new responsibilities in terms of instruction on digital literacy (contradiction D) as well as the need to lobby for students to be given appropriate digital devices (contradiction C), shifts the traditional role of the lecturer (contradiction A) yet affirms the need for lecturers to tailor knowledge for the local context (contradiction B), and requires lecturers to ensure that students use their new resources to learn both theory and application so that they receive a comprehensive Physics education (contradiction E).

A famous paradox states, 'The more things change, the more they stay the same'. This captures the essence of what lecturers are experiencing because they continually have to navigate the way that their role is evolving within an evolving world. Their responsibility as the 'culturally more adept other' does not diminish when new tools become available, because the tools become something more for students to learn. In addition, as Physics resources become more advanced, so too do the roles that students will have to play in the world of work—and it is the responsibility of the lecturers to enable these students to be able to compete in an exponentially more complex world.

References

Ahn, S. (2014). Effect analysis of digital textbook on ability of self-directed learning. In *2014 international conference on soft computing and machine intelligence, 26–27 September* (pp. 135–138). IEEE.

Daniels, H., Edwards, A., Engeström, Y., Gallagher, T., & Ludvigsen, S. R. (Eds.). (2010). *Activity theory in practice: Promoting learning across boundaries and agencies*. Routledge. https://doi.org/10.4324/9780203609439

Gu, X., Wu, B., & Xu, X. (2015). Design, development, and learning in e-textbooks: What we learned and where we are going. *Journal of Computers in Education, 2*(1), 25–41. https://doi.org/10.1007/s40692-014-0023-9

Pepin, B., Gueudet, G., Yerushalmy, M., Trouche, L., & Chazan, D. I. (2015). E-textbooks in/for teaching and learning mathematics: A potentially transformative educational technology. In *Handbook of international research in mathematics education* (pp. 636–661). Routledge.

Chapter 6
Intended Versus Enacted Pedagogy: Technology Triggering Behaviourist Pedagogy in Computer-Based Mathematics Lessons

Abstract The appallingly low results on international benchmarking tests of mathematics obtained by South African children have led to many efforts to intervene in mathematics education in schooling. Especially in the twenty-first century, these interventions favour the use of information communication technology such as computers or iPads. This chapter reports on the contradictions that potentially arise when a novel technology, in this case computer-based mathematics software, is introduced into a mathematics computer-based lesson. Located in a previously disadvantaged school, this empirical study illustrates how a teacher, in an interview, describes his intended pedagogy with computer-based lessons. Making a distinction between intended and enacted pedagogy, the chapter illustrates how in his actual, observed pedagogy, the teacher's intended pedagogy is absent. What happens when a novel tool, such as computer-based mathematics software, is introduced into this teacher's class is that the object of the lesson shifts from mathematical learning to the actual use of the computer. The computer, in fact, becomes the object of the lesson and the outcome is measured in technical skill development, rather than mathematics. Various contradictions emerging in the pedagogy are discussed in this chapter, illustrating how contradictions can function to produce change that is regressive, rather than progressive.

Introduction

South Africa's performance on international benchmarking tests of Mathematics, such as the Trends in International Mathematics and Science Study (TIMSS), is significantly below what one would expect from a middle-income country that spends more, proportionally, on education than larger economies such as the United States (World Bank, 2024). There are many explanations for why children in many South African schools perform poorly on such tests, ranging from underprepared teachers to a lack of material resources (Hoadley, 2017; Hardman, 2023). In classrooms where computers have been introduced as pedagogical tools, there is also a mismatch between teachers' assumptions of how they believe they teach and how

J. Hardman, E. Rzyankina, *Contradictions in Cultural-Historical Activity Theory*, https://doi.org/10.1007/978-3-032-28650-5_6

they do teach (Hardman, 2015). This points to one area requiring intervention if we are going to rectify poor Mathematics performance in South African schools.

This chapter reports on a study of a Grade 6 Mathematics teacher operating in a previously underprivileged school in the Western Cape province of South Africa. This school benefited from a government initiative that placed computers into well-managed underprivileged schools in a bid to improve Mathematics performance. I use CHAT to track the intended pedagogical activity of this teacher versus his actual pedagogy. The findings illustrate that a teacher may believe that learning is an active process yet instead focus on covering the curriculum, adopting a behaviourist rather than a constructivist pedagogy in order to achieve this. While this is a single case study, it is representative of further studies I have been involved in. What it indicates is that there is a primary contradiction in the object of education in schools in South Africa, which is the tension between deep learning and curriculum coverage. This chapter sets out to explain how and why this is the case.

How Technology Can Improve Attainment in Mathematics

Information and communication technologies (ICTs) have been used in many parts of the world in attempts to improve students' performance in Mathematics. The research regarding the impact of ICTs on mathematical attainment clearly points to the fact that ICTs, at a primary-school level, do indeed impact positively on student outcomes (Tamim et al., 2011; Higgins et al., 2012; Chauhan, 2017; Slavin et al., 2009). Drawing on 25 meta-analyses produced over a 40-year period, Tamim et al. (2011) performed a second-order meta-analysis, concluding that 'the average students in classrooms where technology is used will perform 12 percentile points higher' than those in more traditional classrooms where technology is not used (Tamim et al., 2011. p. 17). One meta-study (Campuzano et al., 2009) indicates that mathematics attainment in primary school is not impacted at all by ICTs, but this study is critiqued by Cheung and Slavin (2013) due to what they see as a methodological flaw in the meta-analysis, and they go on to illustrate in their own meta-analysis of 74 studies that attainment in Mathematics is positively impacted using ICTs.

Further findings from the meta-analyses reviewed indicate that students with special needs benefit more from technological input than neurotypical students (Li & Ma, 2010; Higgins et al., 2012) and primary-school students benefit more from ICT use than secondary-school students (Liu et al., 2025; Sabri & Daud, 2014). Gender, race and socioeconomic status do not significantly influence the effect that ICTs have on Mathematics attainment. Moreover, in relation to Mathematics learning, Li and Ma (2010) show that students achieve more positive outcomes when they are tested using non-standardised tests as opposed to traditional standardised tests.

Tamim et al. (2011) further indicate that computer technology that supports instruction is more effective than technology that offers direct instruction, which points to the importance of the pedagogical basis of ICT use. The authors are clear to highlight that technology on its own has little benefit to the student, but how it is designed and how it is used is important in determining whether it will be effective in improving student outcomes. This sentiment is echoed by Higgins et al., who state that it is 'the pedagogy of the application of technology in the classroom which is important: the how rather than the what' (2012, p. 3).

Behaviourist Software

When computer-based mathematics software was initially rolled out in the twenty-first century in schools on a large scale, much of this software was what could be called drill and practice software, underpinned by a Behaviourist notion of learning where students are rewarded for immediate answers and no elaboration of how one arrives at an answer is provided (Kuiper & de Pater-Sneep, 2014). This kind of software (such as one sees in programmes such as Mathletics (see for example, Hardman & Raudzingana, 2021) is premised on the notion that mathematical knowledge needs to be reinforced, rather than focused on the creative development of critical in relation to mathematics. Of course, the need to reinforce knowledge in mathematics is central in grounding a student in mathematics but it is not sufficient to develop a child cognitively in the Vygotskian (1986) sense where learning leads development and is therefore, considered less useful for cognitive development than what is called constructivist software, discussed below.

Constructivist Software

The research into what kind of pedagogy underpins the effective use of technology for student attainment indicates that a constructivist type of pedagogy should form the foundation of teaching with technology (Hardman, 2019; Lam et al., 2021; Diyal, 2024; Rosen & Salomon, 2007). The most popular type of software available for use in schools is termed 'constructivist' software (Britain, 2001) and the most prevalent use of this software plays out in a 'constructivist' environment. However, what exactly is meant in the literature by the word 'constructivism' is somewhat opaque. Historically, this term draws from the notions of Piaget (1976) and Vygotsky (1978) that learning is an active process, requiring participation from the child.

Piagetian theory (1976; 1977) indicates that children learn actively through a process that entails both assimilation, where they understand novel information in terms of pre-existing cognitive functions, and accommodation, where these existing structures shift because the novel knowledge clashes with what is already known. In a teaching scenario, this means that children must be given opportunities to interact with objects to develop their knowledge structures. The process underlying this is referred to as 'cognitive conflict' and it relies on the children being subject to a state of disequilibrium, where previous structures are insufficient to understand new knowledge and must therefore change and grow to accommodate this (Flavell, 2020). Piaget's[1] idea of the role that conflict and disequilibrium play in development speaks to the CHAT notion of contradictions.

Piaget (1977) proposes that cognitive development has to precede learning, and while he sees teaching as important for cognitive development, he does not think it alone ensures that cognitive development and learning will occur. For Piaget (1977) crucial to cognitive development is equilibration, the process whereby disequilibrium is overcome by the child in a new context. Conversely, Vygotsky (1978, 2012) places teaching at the heart of cognitive development, and proposes that learning leads to cognitive development if it is properly organised. As discussed in Chap. 1, for Vygotsky, good learning (which is learning that leads to cognitive development) requires mediation of concepts within a unique social space called the zone of proximal development (ZPD). The primary mediating tool, according to Vygotsky, is semiotic mediation. The idea of a child as an active cognising agent (which both Piaget and Vygotsky share) lends itself to software that allows for creative interaction with bodies of unknown knowledge in order to learn. This requires, in a Vygotskian framework at least, built in mediation in the form of feedback and instruction that can lead a child into their ZPD.

Much research has indicated that Vygotsky's work is empirically sound in relation to teaching and learning in schools (Daniels, 2002; Hedegaard & Chaiklin, 2005; Wells, 1999). I support the view that students in schools should be considered active cognising agents engaged in problem-solving in a mediated context, and in previous work I have defined 'pedagogy' from a Vygotskian perspective. I would like to mobilise this definition in this chapter, stressing that pedagogy is:

> a structured process whereby a culturally more experienced peer or teacher uses cultural tools to mediate or guide a novice into established, relatively stable ways of knowing and being within a particular, institutional context in such a way that the knowledge and skills the novice acquires lead to relatively lasting changes in the novice's behaviour, that is, learning. (Hardman, 2008, p. 69)

The CHAT framework allows me to explore how pedagogy utilising ICTs unfolds in an actual classroom, and by focusing on contradictions in ICT-based lessons I explore the potential for pedagogical change.

In CHAT, contradictions constitute the fundamental mechanism through which activity systems evolve and expand. Engeström views these structural tensions as 'the motive force of change and development' in human activity, representing conflicts between different elements, developmental phases or competing activity systems (1999, 2014). It is their intimate involvement with change that drew us to contradictions as a site of dynamic development.

---

[1] While we refer to Piaget's work in this book, his ontological and epistemological position differs to that of CHAT theorists, although there are many instances where his work and CHAT resonate and can compliment each other.

However, according to the CHAT framework, contradictions cannot be seen until they emerge in language as tensions that require resolution. Engeström's influential discursive manifestations of contradictions framework helps researchers to group contradictions based on how research participants express themselves when they are talking about them (discussed in Chap. 3; see also Engeström & Sannino, 2011; Galleguillos & Borba, 2018).

Engeström's four-level model of contradictions also allows researchers to classify contradictions based on where they occur: within nodes of the activity system, between nodes of the activity system, between time periods or between activity systems (as discussed in Chap. 2). In the sense that contradictions occur at differing levels, they are not unlike ripples in a pond, spreading out from the centre.

Primary contradictions exist within individual nodes of an activity system. In commodity-producing societies primary contradictions often stem from the contrast between a commodity's use value to society and its exchange value to an individual. These contradictions manifest in tensions between quality and quantity, care and efficiency, or local needs and market demands within any single nodes of the activity triangle. Teachers themselves embody primary contradictions, torn between their professional commitment to student development and the economic pressures of their employment conditions. Educational technology reveals similar tensions, as digital tools designed to enhance learning are simultaneously marketed as solutions to reduce educational costs, creating contradictory design imperatives. In a schooling context, the most profound primary contradiction that I have found is one that occurs within the object, between the development of conceptual understanding and curriculum coverage.

A secondary contradiction emerges between different nodes of an activity system. In our context there is a secondary contradiction between the subject, namely the teacher, and the rules, which prescribe a content- and testing-heavy curriculum. This results in a discrepancy between the teacher's belief in an active type of pedagogy (which includes student voices) and the teacher's need to control classroom discourse (in order to quicken the pace of a lesson). A secondary contradiction also occurs when digital technology, part of the mediating artefact, disrupts the object of a lesson, which is to teach a particular subject (as already explored in Chap. 4). In addition, a secondary contradiction can exist between division of labour, namely rigid departmental structures, and the object, interdisciplinary learning objectives. What happens here is that the teacher reverts to traditional chalk and talk pedagogy that sees the object narrow to curriculum coverage rather than deep learning of mathematics (primary contradiction in the object).

Tertiary contradictions arise when a culturally more advanced form of an activity is introduced alongside the dominant form. This often manifests in resistance to change or struggles between old and new ways of working. Technology integration frequently generates tertiary contradictions as schools attempt to adopt digital pedagogies while maintaining traditional classroom management systems, grading practices and spatial arrangements. The introduction of student-centred learning approaches within teacher-centred institutional frameworks creates ongoing tensions between emerging and established educational paradigms.

Finally, quaternary contradictions arise between neighbouring activity systems that share overlapping objects or participants. These meta-contradictions reflect broader systemic conflicts, such as those between educational institutions and workplace demands, or between different professional communities working on similar problems. Further quaternary contradictions exist between schools and other activity systems sharing educational objects. The relationship between schools and families represents a significant source of quaternary contradictions, as each system operates with different objects and rules, which entail different expectations of student development. While schools emphasise standardised curricula and assessment, families may prioritise different forms of learning and development.

The existence of contradictions within and between systems indicates why educational reforms face such challenges in implementation. Facing a contradiction can be difficult and cause unease, with people wanting to operate in the ways they have always operated. However, contradictions can be overcome and new ways of doing and being can emerge through dealing with them.

The first step is to identify the contradictions in the system, which is what this chapter aims to do. The research I report on here was undertaken to clarify variations in pedagogical praxis using technology. The findings point to contradictions at various levels of the activity of teaching with technology, as innovation clashes with tradition.

The Study of ICT Usage in Grade 6 Mathematics Lessons at Sonskyn[2] Primary School

The study explores how the introduction of technology impacts on intended versus enacted pedagogy in the presence of computer-based technology in Grade 6 Mathematics lessons.

The Context of the Study

South Africa has a history of unequal education due to the system of apartheid that systematically undereducated most of the population, who were deemed unable to learn Mathematics and Physical Science to the same level as white students as they were black.[3] (Christie & Collins, 1982; Kallaway, 2002; Hardman, 2023). Although South Africa is now well into a democratic dispensation, the struggle with these subjects continues, with South African children performing very badly in

[2] This is a pseudonym.

[3] The term 'black' is used here as it was used under apartheid to refer to people who were not classified as white and were, therefore, excluded from civil and human rights under apartheid.

international benchmarking tests such as the Trends in International Mathematics and Science Survey (Kanjee et al., 2025). This has led to various interventions in schools aimed at improving attainment in Mathematics and those subjects incorporating 'physical science' (Life Skills in the Foundation Phase, Natural Sciences in the Intermediate and Senior Phases, and Physical Sciences in the Further Education Phase).

The Research Design

The study sought to understand whether there were underlying contradictions between a teacher's intended pedagogy and his enacted pedagogy in the presence of novel technology. To achieve this, I adopted a case-study design for data collection, selecting a teacher and his pedagogical praxis as a single case study (Yin, 1984; Denzin & Lincoln, 2005). As discussed earlier, the theoretical framework underpinning this study is CHAT, which remains, I argue, underrepresented as a framework for studying pedagogy in mathematics classrooms (although see for example Tomaz & David, 2021). The case investigated in this chapter was purposively selected from a group of schools that benefited from a government initiative in the Western Cape province of South Africa, placing computers with mathematical software into classrooms. I chose to focus on Grade 6 in this study, understanding that children at this stage are developmentally able to think concretely through problems but are also on the cusp of more abstract reasoning (Piaget, 1976). Grade 6, in South Africa, is referred to as the Intermediate Phase (IP) and has as its focus reading to learn, rather than the earlier Foundation Phase (discussed in Chap. 7) which focuses on learning to read. The object of the IP is to develop school-based concepts in children.

Data Collection

Data collection was carried out by means of an interview with the teacher (see Chap. 3, Table 3.1 for the outline of this interview) and through the collection of video data that was transcribed and analysed in terms of the schedule discussed in Table 3.2 in Chap. 3. Both the interview and the observation schedule use CHAT deductively to map data onto the various nodes of the activity system triangle.

The Teacher

Mr. B is 37 years old and has been teaching for 13 years. He holds a higher certificate in education from a teachers' training college. He is a first-language Afrikaans speaker, and he teaches in Afrikaans. While he has taught different subjects in the past, he has been teaching mathematics across Grades 6 and 7 for the past 5 years.

The School

Sonskyn Primary School is in a farming district 70 kilometres outside of Cape Town. The school is small, with 7 teachers and 143 students. The school building is made of brick and mortar as well as straw bale. It charges no school fees and is classified as a Quintile 1 school.

South African schools are separated into 5 quintiles, with Quintile 5 consisting of the most well-resourced, highest fee-paying schools, and Quintile 1 consisting of the most poorly resourced, no-fee schools. Quintile 1 schools represent the poorest 20% of schools, and are typically located in rural and impoverished communities, while Quintile 5 schools serve the least poor 20% of learners in the most affluent areas and are supported by means of school fees from parents or guardians (White & Van Dyk, 2019). Despite a progressive funding model where Quintile 1 schools receive significantly more government funding per learner than their Quintile 5 counterparts, the former remain significantly worse off in terms of school resources and overall composition (Maistry & Africa, 2020). This disparity reflects the broader systemic challenges within South African education, where historical inequalities and socioeconomic factors continue to perpetuate educational disadvantage despite compensatory funding mechanisms designed to address these imbalances.

The school has seven teachers, a secretary and a groundskeeper. The school receives a grant from the government to provide children with a meal at lunchtime. This meal consists of two slices of bread with jam and peanut butter, and a cup of a vitamin-enriched drink currently used in the nutritional management of HIV/AIDS patients. The principal and the teachers seek to instil notions of sharing and acceptance in the children, promoting a culture of fairness and tolerance within the school. So, for example, on one of my visits the principal suggested that those children lucky enough to have school uniforms and shoes should not tease children who did not have them. As temperatures drop to below zero degrees Celsius in winter, the principal has managed to acquire shoes for the children from a generous donor. In theory, the school prescribes the wearing of full school uniform, but most children are unable to afford a uniform.

Interviews with teachers and the principal suggest that the school has a good relationship with the community, and the surrounding farms have been known to provide the school with barrels of apples, the major crop in the area, to help them raise funds. Parents are described by the principal as ‘very involved’ in the school and they are encouraged to attend computer lessons after school hours so that they too can become computer literate.

In the classroom, students sit in desks of groups of four, facing each other, rather than facing the blackboard. According to Mr. B this seating arrangement aims to ‘promote group work’. When students work in the computer laboratory they work in self-selected friendship pairs. Below the findings of the study are elaborated.

The Subject

Teachers' epistemic assumptions regarding how students learn are likely to impact on their pedagogical practices. Teachers' perceptions of how to teach Mathematics influence both what tools they select as well as how they use these tools. Therefore I began my interview with Mr. B by probing for his assumptions about how children learn, and he explained:

> I know that children learn by doing stuff… Right let's take for example, um, times table how do we do it. Right we gonna count in twos so you use your whole body standing up two four six eight ten [demonstrates movements with his hands] … and then [to consolidate learning] he's got to do an exercise with his friend they got to work something else out, how can they come, but as they playing they also learning from each other…

Mr. B believes that children learn through actively engaging in problem-solving, through collaborating with a peer and through play. This engagement often has a kinaesthetic element, such as actively using your body as a tool for learning to count. The notion of learning through play that is articulated by Mr. B is suggestive of low levels of teacher control over the pace of a lesson, because play and exploration are student driven. The slowing of pace and recognition of active students is suggestive of a constructivist pedagogy.

However, when asked about whether his teaching changes in the computer-based lessons, Mr. B indicates that in these lessons 'they must copy my actions on the computer, exactly, exactly'. The notion of copying that he articulates is suggestive of high levels of teacher control over the pace of a lesson, and is suggestive of a behaviourist pedagogy.

The Mediating Artefact

When I asked what tools he used in a computer-based lesson, Mr. B indicated that he used the computer as a tool to practice Mathematics, as shown in Extract 6.1.

Extract 6.1: Tool Use

1. **Mr B:** Uh, I can say that yes, it is the computer that is the tool. I don't use the blackboard.
2. **Interviewer:** The tool for?
3. **Mr B:** To practise maths. It is very good for practice.

However, if we look at how the computer is actually used in the computer-based lesson, it becomes clear that teaching the technology becomes the object of Mr. B's lesson. While he intends to enact both a Mathematics drill-and-practice and constructivist session with computers, in his computer-based lessons he actually instructs students how to use computer software using language. This is illustrated in Extract 6.2.

Extract 6.2: Language as the Mediating Artefact Versus the Computer as the Mediating Artefact

1. **Mr B:** You click on it. *Stands behind Ephraim.*
2. Click on it. *Points at the screen; points at the icon on the screen.*
3. Alt. *Points to the 'alt' button; then presses the alt button.*
4. Alt.
5. It must be blue the 'icon' that one clicks on must be blue in order for one to click on it and work with it.
6. Remember, if it's blue you can work with it. *Looks at his watch.*
7. That's it! *Looking at Joe's work.*
8. Look! *To Nkosi, who is unable to find the paint function.*
9. Look now! *Points to it with his finger.*
10. Look! *Takes Nkosi's mouse and moves it to the required place, speaking as he does so.*
11. At the bottom [of the screen] *Gives the mouse back to Nkosi and walks away.*
12. At the bottom.
13. Faster…
14. *Choep.*[4] *Waving his hand and putting his finger on his lips to indicate silence is required.*
15. *Choep. Walks around the classroom.*

In the above extract, the teacher's language is predominantly related to directing the students' task-oriented actions in relation to technological literacy.

Rules

The rules in the system are the norms that guide action and interaction. They could be at a macro level, such as the curriculum document that prescribes what is taught, or at the micro level, such as the teacher and student agreeing that students must not call out in class but must put up their hand. I suggest two types of rules are distinguishable:

- Instructional rules, which incorporate both rules of pacing (analysed in terms of levels of teacher control) and rules of evaluation (analysed in terms of whether the teacher elaborates the criteria required to generate an appropriate answer)
- Social-order rules, which relate to behavioural prescriptions and proscriptions (analysed in terms of teacher control over these rules).

---

[4] This is a shortening of an Afrikaans phrase: '*Choep stil*', which means to keep 'dead quiet'. I have not translated this because the use of '*choep*' in the extract brings across the sense of control over disciplinary rules more readily than the English translation would.

While rules can be both linguistic and non-linguistic, the nature of the data presented in this thesis is predominantly linguistic and what follows is an investigation of linguistically encoded rules.

In relation to how he anticipates teaching in the computer-based lesson, Mr. B elaborated on how he slows the pace so that all the students can keep up. He explained:

> If you, if you do um a lesson like this you've got to do it as a group. Because if I say now go and take out 'equation editor' to everyone they not that, all of them are not that quick with the mouse or at the same level, which is understandable.

While this statement shows that he intended to slow the pace to include slower students, Extract 6.2 shows that in his enacted pedagogy he tried to speed them up. In lines 13–15, Mr. B tightly controls the pacing.

In terms of rules of evaluation (where the teacher explains what counts as a meaningful answer), Mr. B claims that, 'I will explain, you know, go over it. Go back. Put it all out there. Even, uhm, today I used that apple to show, look I cut him in half. Then they seeing that.' However, in his actual practice in the computer-based lesson the rules of evaluation are not elaborated, as we see in Extract 6.3.

Extract 6.3: Enacted Pedagogy in the Computer-based Lesson

1. **Mohammed:** Sir? *Pointing at the screen.*
2. **Mr B:** You press shift, and then the sign.
3. Shift and then… *Looks at the screen and stops talking as he realises that it is not a technical issue that the student has but a conceptual one. Mohammed does not know how to compare fractions of different sizes.*
4. 1¼. Is it bigger or smaller than 1?
5. Bigger or smaller?
6. **Mohammed:** Smaller.
7. **Mr B:** Huh? *Looks at his watch.*
8. **Mohammed:** Bigger?
9. **Mr B**: Hmm, shift and bigger. *Rather than elaborating the answer and showing how to compare the fractions, the teacher takes the mouse and selects the right answer before walking away to check on another student's work. Looks at his watch.*
10. **Mr B:** Shew! People, time is short!
11. Everybody finished? *Does not wait for a response to the question.*
12. Everybody finished?
13. Quickly.
14. Quickly.
15. **Jakob**: Sir? *Points to his computer screen, on which he has drawn a shape, as requested. The second part of the task requires that he colour half of the shape in; he has not done this and hence, puts up his hand to ask for assistance.*
16. **Mr B**: We not doing that now.
17. We moving on.
18. Next one.

19. Next one, go to 'Menu'.
20. All together.

In this extract we can see Mohammed asking a question; the teacher seems to think the question being asked is technical and therefore tries to help him with the computer. However, the teacher then realises that Mohammed is struggling to understand how to determine which fraction is bigger or smaller. The teacher gives the answer here, without giving reasons for why this counts as a correct answer. This indicates the evaluative rules are opaque in this context.

We can also see rules of social order here, with the teacher in lines 15–20 very explicitly controlling the selection and sequence of the lesson's content, and getting the students to move through the work as a group.

Object

The object of an activity is the motive for that activity. In schooling, you would assume that the object of most teachers' pedagogy would be teaching a subject's content, concepts and procedures so that the students learn these.

Yet in relation to the object of the computer-based Mathematics lessons, Mr. B indicated that it is the teaching of computer skills *and* mathematical understanding that motivates the activity:

> I saw that you've got to get kids to a certain level where they are computer literate… So they have to learn how to use the computer, and, I would say, ja, that I have to teach them computer skills as well as maths in this lesson.

For Mr. B, rather than mathematical understanding being the sole object of a computer-based Mathematics lesson, the technological tool itself becomes the object. Teacher instructions become the tool used to achieve this object. When asked to comment on the object of the computer-based Mathematics lesson, Mr. B indicated that it was concerned with developing mathematical understanding as well as computer skills, but this is contradicted by the evidence from the observations, which shows that teaching students how to use the software became the only object. (This situation is similar to that described in Chap. 4.)

In Extract 6.4, we see how Mr. B focuses on developing the skills of a particular student.

Extract 6.4: Computer Skills as the Object

1. **Mr B**: OK, Anil, *Talking only to Anil.*
2. Look at the screen. *Mr B is on his own computer. What he types on his screen is displayed on his screen.*
3. You going to click there.
4. OK, up there, you see in the blue?
5. Or double click.
6. Double click.

7. Let me show you. *Mr B goes to Anil's computer and Anil stands up so Mr B can navigate the screen for him.*

In this extract Mr. B uses spoken language and the computer to develop the student's technical skill in a specific task. This is a localised context-bound object with a specific outcome: a technologically literate student. Anil's subsequent engagement with the computer in the rest of the episode points to the development of greater technological literacy but says nothing about his grasp of mathematics.

Mr. B does perceive that the computers could hinder the students' learning of Mathematics, but he emphasises their helping role in Extract 6.5.

Extract 6.5: Motivational Aspects of Using Computers
Mr B: … it is actually a tool that we can use to, to it's like a carrot keeping it in front (dangles his pen in front of his face)

Interviewer: It's motivational?

Mr B: Ja.

For these students, simply being able to use a computer encourages engagement. None of the students in this school have access to computers in their own homes, so the opportunity to use this technology can indeed be motivating, although only extrinsically. This extrinsic motivation afforded by the computer-based tasks is an important point to recognise as learning mathematics requires, as with all abstract concepts, the motivation to engage with the concepts under discussion. if one could harness the technology in such a way that it could not only motivate, but also mediate, engagement, this would prove useful to learning mathematics. Steps in this direction have been taken up worldwide with the introduction of software that is explicitly designed to mediate engagement in an active manner such as GoeGebra (Birgin & Topuz, 2021). The extent to which this is succeeding continues to be studied, but there is emerging evidence that this type of software is underpinned by a constructivist pedagogy (Hundeb, 2024; Seloane et al., 2023).

The Community

In CHAT the community includes only those people who share the object. It does not refer to a wider community that is not itself engaged in, in this instance, developing mathematical understanding. The community of both the intended pedagogy and the enacted pedagogy is made up of the students and the teacher. There are no laboratory assistants or assistant teachers in this school and no parents volunteer to assist in the classroom.

However, Mr. B does indicate in his interview that parents are allowed to come and use the computers on Saturdays where they are given training in basic computer skills by himself and a staff member. It is worth noting that none of the parent body of this Grade 6 class has finished school with a matriculation certificate, which would show that they had passed the examinations at the end of their schooling career. They are mostly farm labourers, although some are unemployed and one father is a tractor driver. Note, however, that in a CHAT sense, the parents do not share the object of the mathematics lesson, as far as can be ascertained from interviews and observations.

Division of Labour

In this study, division of labour is analysed by focusing on the roles of the teacher and the students, and the shifts in the relations between these. I do this by looking at the data in terms of both linguistic and non-linguistic categories. When we view the linguistic data, we can examine the degree of the asymmetrical power relations that exist between the teacher and students, because the extent of this traditional structure can shift when students' voices are encouraged or when students insist on making themselves heard. When we view the non-linguistic data, we can look at whether there are firm boundaries between the teaching and learning spaces. Where firm boundaries exist, there is likely to be a high degree of asymmetrical power, with the teacher occupying a more traditional didactic role. Where the boundaries between teaching and learning spaces are more fluid, you can expect to see a shift in roles in the lesson.

When asked what role he takes when teaching, Mr. B responded:

> Well, I must know the mathematics but I must also allow them to talk to, I would say, to be in discussion about what we are doing. To ask questions and that kind of thing. They are not just these empty buckets; they can bring their own ideas.

This shows that Mr. B. sees himself as a mediator of students' active engagement with the task under discussion. Mr. B thinks children learn actively, through discussion, with tools to understand mathematical concepts in a context where the rules of pacing are lightly controlled by the teacher and division of labour has the teacher acting as a mediator and the students acting as active cognising agents. Again, this is suggestive of the type of pedagogy generally referred to as 'constructivist',[5] which views children as active cognising agents, who co-construct knowledge through the mediation of a more competent other.

This picture changes slightly when he speaks about the computer-based lessons where the actual enacted object very often is merely teaching the skills required to use the computer, although the stated desired object is still to teach mathematical understanding. In the actual computer-based lesson, the evidence shows that the

[5] We are aware that the term 'constructivism' has become so widely used that it has led to a general conceptual slippage around what the pedagogy implied here entails. For us, constructivist pedagogy is informed by the work of Vygotsky (2012) where mediation in the zone of proximal development serves a developmental goal. It is also true that constructivist pedagogy draws on the work of Piaget (1976), in the sense that it recognises that children are active cognising agents. However, Piaget did not espouse a theory of pedagogy, but rather, a theory of development, which is why his work is not dealt with in any depth in this chapter.

division of labour becomes far more asymmetrical than Mr. B's interview would suggest, as seen in Extract 6.6.

Extract 6.6: Division of Labour as Enacted in a Computer-based Lesson

1. **Mr B:** Everyone must be on the second shape now.
2. Everyone has drawn the square.
3. Now draw a rectangle.
4. All together.
5. No talking please.
6. It's just polite.
7. Are you listening?
8. Off! *Raises his voice slightly.*
9. Quickly, quickly.
10. Time is short.
11. Time is very short.

There is a very asymmetrical relationship between the teacher and taught in this computer-based lesson. The teacher is completely in control of the rules of engagement (lines 5 and 6 indicate no talking is allowed) and he directs students to complete the task in time. The role that the teacher enacts here is one of director, in that he directs behaviour and interaction completely. There seems little to no space for student voices here.

This suggests that a very different pedagogy occurs between what is intended and what is enacted in a computer-based Mathematics lesson in Grade 6. Figures 6.1 and 6.2 show the two distinct systems, enabling us to see whether there are contradictions that have emerged and whether these may point to changes in the systems.

Figure 6.1 represents Mr. B's intended pedagogy that was elicited during the interview. What we can see is that he has a belief that children learn actively and that they should be engaged in knowledge acquisition. He uses a variety of tools in his classroom, ranging from the blackboard to various props he may bring to class. However, his most used tool is language, which he uses to develop students understanding of mathematics. The outcome of his pedagogy is mathematically sound students. Social order rules are implicit, and he believes that pacing should be informed by students' progress. In terms of evaluative rules, Mr. B believes that the response regarding what counts as a valid answer to a question should be elaborated and well explained, with examples where feasible. Regarding division of labour, Mr. B conceiving of himself as a mediator and students as active cognising agents. The picture we have in Fig. 6.1 is the ideal intended pedagogy as expressed by Mr. B. What really happens in the classroom when technology is used, however, shifts this picture.

Figure 6.2 represents Mr. B's enacted pedagogy enacted in computer-based Mathematics lessons that were observed, videotaped and transcribed. The circled nodes show where the primary contradictions occur and the double-headed arrows show where the secondary and tertiary contradictions occur.

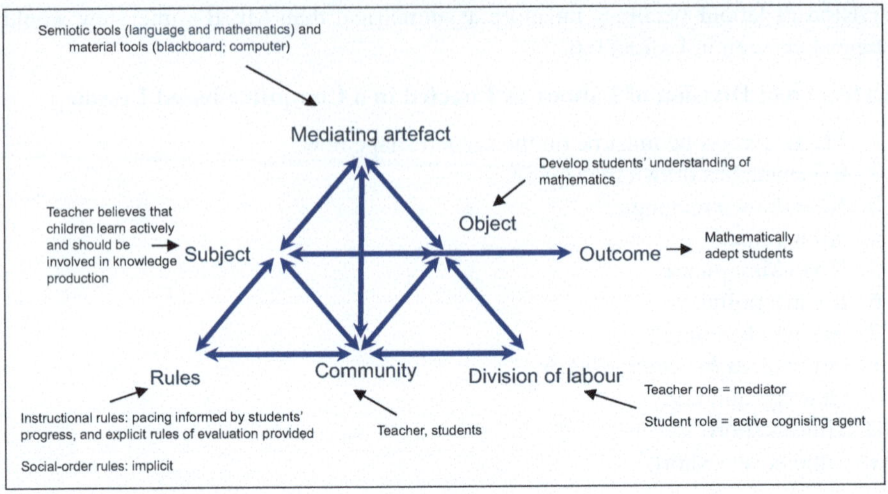

Fig. 6.1 Constructivist pedagogy as intended pedagogy

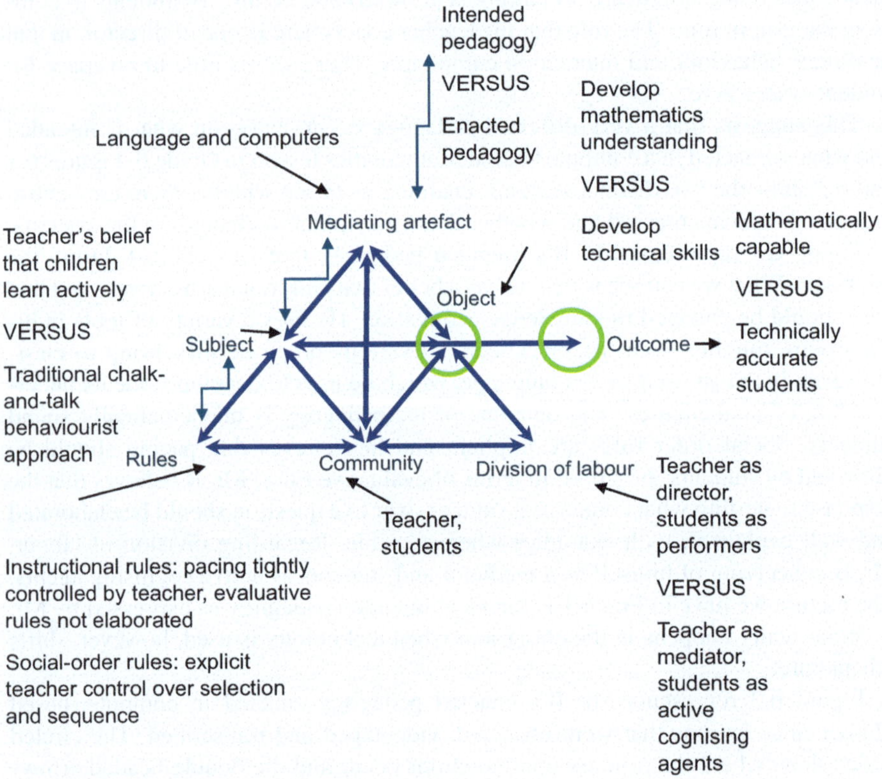

Fig. 6.2 Directive pedagogy as enacted pedagogy

In the object, we can see a primary contradiction emerging between the development of mathematical understanding on the one hand, and the need to teach technical computer skills on the other hand. This dissonance in the object is also reflected as a primary contradiction in the outcome of the activity system, with mathematical ability and computer skills and apparently co-existing. Another primary contradiction emerges in the division of labour between the teacher's stated role as mediator versus his actual role of maintaining the direction of behaviour through strong pacing and tight control over sequencing and selection of work. This means that the agentic roles that he envisioned for his students are reduced to passive roles. We may well ask, what object will gain primacy in this system, mathematical understanding or computer-based skills? These are, after all, Mathematics lessons.

Our answer lies in the secondary contradictions in the system, between the nodes of subject and rules and between the nodes of mediating artefact and object. In this study, the teacher's stated desire to teach in what we have called an active constructivist manner is in opposition to the rules of the activity of teaching Mathematics in a computer-based classroom. Change happens when this contradiction surfaces. We can see the teacher then tightens control over pacing, sequencing and selection of content, moving away from a view of students as active cognising agents to students as passive recipients of knowledge and, in this instance, narrowing the object of the lesson to technical computer skills. The contradiction between using the computer as a tool to it becoming the object of the lesson emerges from a secondary contradiction between the mediating artefact and the object of the lesson.

Finally, we can note a tertiary contradiction between the intended pedagogical model and the enacted model. Rather than moving to a new way of teaching with technology, Mr. B reverts to a traditional chalk-and-talk and extremely tightly controlled lesson in the computer-based Mathematics lesson. What we can see then, from a comparison between intended and enacted pedagogy is that contradictions in a system can be both progressive, leading to an expansion of learning, or regressive, leading to a narrowing of what is taught in a lesson. What we see in the picture presented here is a regression to older forms of pedagogy rather than a new pedagogy emerging from the use of new technology. (The fact that systems can regress in response to the contradictions is discussed in Chap. 8.) This situation is not one that we can locate in the teacher, which is why a CHAT analysis enables us to see the various stressors placed on him due to the introduction of novel technology. in his interview Mr. B. indicated that he had had only 1 h of training in the technology and that he himself did not possess a personal computer. The added burden of a loaded curriculum to cover, sees him trying to manage a novel situation as best he can, which means he reverts to traditional, tightly paced lessons in the computer-based lesson. Without a CHAT analysis, it would be impossible to see the nuance behind the activity of teaching in a previously disadvantaged school with novel technology.

Conclusion

There is great hope that novel technology in the twenty-first century can lead to better outcomes in classrooms, especially in subjects such as Mathematics. The current chapter looked at how a South African teacher in an underprivileged school described his intended pedagogy with computers versus his enacted pedagogy with computers. Findings indicate that, while the teacher intended to use the technology as a tool to develop students' mathematical understanding, what happened was that the object of the Mathematics lesson instead became developing students' computer skills. As a ripple effect from this primary contradiction, rather than moving to a novel pedagogy with technology, the teacher reverted to a traditional mode of teaching where very little agency remained in the students' roles. This illustrates that contradictions are not necessarily productive of novelty or change, but can, in fact, lead a system to revert to a previous version of that activity. Can these contradictions be overcome in a way that moves the activity forward? There is indeed a way to develop Mr. B's pedagogy in relation to computer-based lessons, and this would be through in-service training, so that he becomes familiar with the technology. The fact that the computer-based lessons elicited the motivation to engage with the technology speaks also to the possibility of the technology being refined in such a way that it harnesses motivation in relation not solely to technology, but to the mathematical concepts being covered.

References

Birgin, O., & Topuz, F. (2021). Effect of the GeoGebra software-supported collaborative learning environment on seventh-grade students' geometry achievement, retention and attitudes. *Journal of Education in Science, Environment and Health, 7*(4), 282–299.

Britain, G. (2001). ImpacT2: Emerging findings from the evaluation of the impact of information and communication technologies on pupil attainment.

Campuzano, L., Dynarski, M., Agodini, R., & Rall, K. (2009). *Effectiveness of reading and mathematics software products: Findings from two student cohorts*. NCEE 2009-4041. National Center for Education Evaluation and Regional Assistance.

Chauhan, S. (2017). A meta-analysis of the impact of technology on learning effectiveness of elementary students. *Computers & Education, 105*, 14–30.

Cheung, A. C., & Slavin, R. E. (2013). The effectiveness of educational technology applications for enhancing mathematics achievement in K-12 classrooms: A meta-analysis. *Educational Research Review, 9*, 88–113.

Christie, P., & Collins, C. (1982). Bantu education: Apartheid ideology or labour reproduction? *Comparative Education, 18*(1), 59–75. https://doi.org/10.1080/0305006820180106

Daniels, H. (2002). *Vygotsky and pedagogy*. Routledge.

Denzin, N. K., & Lincoln, Y. S. (Eds.). (2005). *The SAGE handbook of qualitative research* (3rd ed.). Sage Publications.

Diyal, S. B. (2024). Integration of Constructivist Pedagogy and ICT Tools to Enhance Learning Outcomes. *Educational Journal, 3*(1), 59–67.

Engeström, Y. (1999). Activity theory and individual. *Perspectives on activity theory*, 19.

Engeström, Y. (2014). *Learning by expanding: An activity-theoretical approach to developmental research*. Cambridge University Press.

Engeström, Y., & Sannino, A. (2011). Discursive manifestations of contradictions in organizational change efforts: A methodological framework. *Journal of Organizational Change Management, 24*(3), 368–387. https://doi.org/10.1080/10508406.2016.1204547

Flavell, J. H. (2020). *The developmental psychology of Jean Piaget*. Barakaldo Books.

Galleguillos, J., & Borba, M. D. C. (2018). Expansive movements in the development of mathematical modeling: Analysis from an activity theory perspective. *ZDM, 50*(1), 129–142.

Hardman, J. (2008). Researching pedagogy: An activity theory approach. *Journal of Education, 45*(1), 65–95.

Hardman, J. (2015). Pedagogical variation with computers in mathematics classrooms: A cultural historical activity theory analysis. *PINS, 48*, 47–76. ISSN 2309-8708.

Hardman, J. (2019). Towards a pedagogical model of teaching with ICTs for mathematics attainment in primary school: A review of studies 2008–2018. *Heliyon, 5*(5).

Hardman, J. (2023). *A cultural-historical approach towards pedagogical transitions: Transitions in post-apartheid South Africa*. Bloomsbury. ISBN 978-1-3501-6470-3.

Hardman, J., & Raudzingana, M. (2021). Mathletics software and student attainment in grade 4-a cultural historical analysis. *Advances in Social Sciences Research Journal, 8*(5), 517–531.

Hedegaard, M., & Chaiklin, S. (2005). *Radical-local teaching and learning: A cultural-historical approach*. Aarhus University Press.

Higgins, S., Xiao, Z., & Katsipataki, M. (2012). *The impact of digital technology on learning: A summary for the education endowment foundation*. Full Report. Education Endowment Foundation.

Hoadley, U. (2017). *Pedagogy in poverty: Lessons from twenty years of curriculum reform in South Africa*. Routledge.

Hundeb, A. B. (2024). Effects of GeoGebra-assisted instructional methods on students' conceptual understanding of geometry. *Cogent Education, 11*(1), 2379745. https://doi.org/10.1080/2331186X.2024.2379745

Kallaway, P. (Ed.). (2002). *The history of education under apartheid, 1948–1994: The doors of learning and culture shall be opened*. Pearson South Africa.

Kanjee, A., Chetty, M., Shozi, N., Ahmed, A., & Correa, F. (2025). *TIMSS 2023: Highlights report South Africa*. Department of Basic Education. https://www.education.gov.za/Portals/0/Documents/Reports/2024/TIMSS%202024.pdf

Kuiper, E., & de Pater-Sneep, M. (2014). Student perceptions of drill-and-practice mathematics software in primary education. *Mathematics Education Research Journal, 26*(2), 215–236. https://doi.org/10.1007/s13394-013-0088-1

Lam, P. L., Ng, H. K., Tse, A. H., Lu, M., & Wong, B. Y. (2021). eLearning technology and the advancement of practical constructivist pedagogies: Illustrations from classroom observations. *Education and Information Technologies, 26*(1), 89–101.

Li, Y., Liu, X., & Wang, Y. (2025). The impact of information and communication technology (ICT) on learning outcomes in early childhood and primary education: A meta-analysis of moderating factors. *Frontiers in Psychology, 16*, 1540169. https://doi.org/10.3389/fpsyg.2025.1540169

Li, Q., & Ma, X. (2010). A meta-analysis of the effects of computer technology on school students' mathematics learning. *Educational Psychology Review, 22*(3), 215–243.

Maistry, S. M., & Africa, I. E. (2020). Neoliberal stratification: The confounding effect of the school poverty quintile ranking system in South Africa. *South African Journal of Education, 40*(4), 1–9.

Piaget, J. (1976). *The language and thought of the child* (M. Gabain, Trans., 3rd ed.). Routledge & Kegan Paul. (Original work published 1923).

Piaget, J. (1977). *The development of thought: Equilibration of cognitive structures* (A. Rosin, Trans.). Viking Press. (Original work published 1975).

Rosen, Y., & Salomon, G. (2007). The differential learning achievements of constructivist technology-intensive learning environments as compared with traditional ones: A meta-analysis. *Journal of Educational Computing Research, 36*(1), 1–14.

Sabri, N. M., & Hj Daud, S. M. (2014). Effectiveness of ICT integration in schools: A survey in Malaysian schools. *International Journal of Education and Development using ICT, 10*(2), 77–94. https://files.eric.ed.gov/fulltext/EJ1105224.pdf

Seloane, O. M., et al. (2023). Developing undergraduate engineering mathematics students' conceptual and procedural knowledge of complex numbers using GeoGebra. *Pythagoras, 44*(1), a763. https://doi.org/10.4102/pythagoras.v44i1.763

Slavin, R. E., Lake, C., & Groff, C. (2009). Effective programs in middle and high school mathematics: A best-evidence synthesis. *Review of Educational Research, 79*(2), 839–911.

Tamim, R. M., Bernard, R. M., Borokhovski, E., Abrami, P. C., & Schmid, R. F. (2011). What forty years of research says about the impact of technology on learning: A second-order meta-analysis and validation study. *Review of Educational Research, 81*(1), 4–28.

Tomaz, V. S., & David, M. M. (2021). Transfer of learning as boundary crossing between cultural-historical activity systems. In *Transfer of learning: Progressive perspectives for mathematics education and related fields* (pp. 221–249). Springer International Publishing.

Vygotsky, L. S. (1978). *Mind in society: The development of higher psychological processes* (Vol. 86). Harvard University Press.

Vygotsky, L. S. (2012). *Thought and language* (Vol. 29). MIT Press.

Wells, G. (1999). *Dialogic inquiry* (pp. 137–141). Cambridge Press.

White, C. J., & Van Dyk, H. (2019). Theory and practice of the quintile ranking of schools in South Africa: A financial management perspective. *South African Journal of Education, 39*(Supplement 1), s1–19.

World Bank. (2024). *Government expenditure on education, total (% of GDP)* [data set]. https://data.worldbank.org/indicator/SE.XPD.TOTL.GD.ZS

Yin, R. K. (1984). *Case study research: Design and methods*. Sage Publications.

Chapter 7
A Dialectical Model of Pedagogy: Beginning Knowledge as a Site of Cognitive Development

Abstract Access to and engagement with science knowledge is key to successfully negotiating the world of the twenty-first century. Research over decades shows that the subject of science, and in particular, the ability to use the scientific method, underpins success in critical and creative thinking. However, in South Africa, attainment in science in schools is extremely low. Many reasons are given for this ranging from lack of material and human resources, teachers' lack of scientific literacy, and large classes. However, perhaps one of the most significant mediators of science conceptual development is the pedagogy that teachers use to teach science. This chapter deals with the question of how science is taught in grade 2 and proposes a model of pedagogy that is geared towards inclusive development. The study reported in this chapter draws on observations of teaching from two teachers, one taught to use the novel method outlined in the chapter, and the other not taught in this method. Findings indicate that the novel method of pedagogy facilitates more dialogue between teacher and student, indicative of more inclusion of student voices, considered key to learning.

Introduction

This chapter outlines a model of pedagogy that we argue can develop a student. Specifically, we attempt to illustrate how a teacher can use a dialectical method to teach students in schools, which are institutions that allow little by way of change. This chapter is empirically located in South Africa, a country with a legacy of unequal education based on racial differences articulated in the apartheid laws. Thirty-one years into the new democratic dispensation, South Africa continues to struggle with education, having the lowest educational outcomes on the African continent (DBE, 2024; Kanjee et al., 2025). The reasons for South Africa's continued educational crisis are multifaceted, ranging from the impact of poverty on schooling to underprepared teachers. In this chapter we deal with the element of schooling that is utterly essential for students' development: teaching.

J. Hardman, E. Rzyankina, *Contradictions in Cultural-Historical Activity Theory*, https://doi.org/10.1007/978-3-032-28650-5_7

The Social Context

Research indicates that attending an institute of higher education and obtaining a qualification is the best way to ensure that one has a chance of obtaining employment, which is especially significant in a country such as South Africa with a 45.5% youth unemployment rate (Ashby-King & Anderson, 2022). South African students sit a final examination at school in Grade 12, and it is this final mark that determines whether the student will be accepted into tertiary education. If students are not taught in a manner that develops them, there can be little hope that they will pass the final school examinations with sufficient marks to enter tertiary education, perpetuating a cycle of poverty and injustice in access to knowledge.

The current pedagogical praxis in South Africa, notwithstanding numerous interventions to alter this, is still largely based on a behaviourist chalk-and-talk model, which may be useful for making students remember information but serves no purpose in conceptual development (Hoadley, 2017). The question then becomes: what kind of pedagogy is optimal for development? We turn to the work of Vygotsky (1978, 2012) to provide us with a constructivist theory of pedagogy that is developmental and transformative. This chapter concerns itself with the following questions:

1. What does an inclusive developmental pedagogy look like?
2. What does this kind of pedagogy achieve in a classroom that traditional pedagogy may not achieve?

The Concepts

Vygotsky argues that all psychological functions begin as real relations, externally, before being turned inwards to form higher cognitive functions.[1] For Vygotsky, the only way to develop is through the mediation of a culturally more competent other in the ZPD (see Chap. 1). It is in this uniquely social zone, opened between the more knowledgeable other and the developing child, that development happens through learning. Vygotsky held the view that learning led development and his theory was very explicitly one of development, not learning in the narrow sense of content acquisition. For Vygotsky, the object of mediated action in the ZPD is conceptual, not technical. In this conceptual framework, teaching and learning are dialectically entailed; Vygostsky uses the term '*obuchenie*', which means something like 'teaching/learning' and refers to a pedagogical praxis that is inclusive of both teacher and taught.

The object of this kind of developmental teaching is geared toward the acquisition of higher cognitive functions, through the interpenetration of scientific and

[1]This is not to assert that biology has no role in development; rather, higher cognitive functions, what we might call executive functions, are entirely social in aetiology.

everyday concepts. It is here in the co-construction of knowledge in the ZPD that a dialectical pedagogy can be situated, as the voices of the students are not marginalised but rather harnessed in the meaning-making interaction in classrooms.[2] While everyday concepts are filled with common sense, scientific (abstract) concepts contain meaning that is shared in a cultural community. When they are not related to everyday concepts, scientific concepts are meaningless and lack relevance in student's life. The object of learning, then, contains both sense and meaning, where 'sense' refers to the idiosyncratic sense a student makes of the object and 'meaning' refers to the shared understanding of the object. In her use of the double move in education, Hedegaard (1998, Hedegaard & Edwards, 2023) illustrates how abstraction and the everyday are interconnected through the teacher's use of general concepts by drawing on situated problems that are meaningful to the students' lived experiences and their developmental stage. Practical application allows scientific concepts to have meaning. Essentially the double-move is a dialectical approach to teaching where the focus is on the teaching activity moving between the general abstraction being taught and situation specific applications. The double-move involves the teacher guiding learning activity from two perspectives: general concepts and engaging students in situated problems that are meaningful to their developmental stage and life situations.

The Pedagogical Model: Teach-Assess-Teach

The pedagogical model presented here draws on Vygotsky's mediation (2012), Hedegaard's (1998) double move and Craig and colleagues' work in the 1990s on the Teach-Test-Teach (TTT) initiative introduced at the then Natal University as an alternative access route to students who did not qualify for direct admittance to tertiary studies.

The kind of development envisaged by the TTT programme involved a student's ability to reflect on their learning metacognitively. For Craig (1996) this requires that a student can provide evidence for arguments they make, rather than beliefs or opinions, which entails that they recognise that evidence has a meaning that is fixed within disciplinary boundaries. The ability to achieve this means that students must be able to step outside of their everyday experiences to deal with novelty. In the TTT programme, the teacher provokes active engagement from students by presenting them with a novel task to which the answers are not generally known by the students, causing cognitive dissonance or conflict (a manifestation of a contradiction between what is known and what is unknown). Assessment functions as a mid-point in teaching and informs teaching.

[2] This is not to suggest that students and teachers share the same content knowledge. Teachers have been taught concepts that students have yet to acquire and they have been embedded in certain epistemic traditions that fix meanings (not for always, but until novel knowledge is achieved).

While the TTT programme is geared to facilitate access to tertiary studies, the Teach-Assess-Teach (TAT) model outlined below is geared towards teaching and learning in schools, and draws on the double move situated within an ethics of care, making it distinct from the TTT programme. Figure 7.1 below represents this pedagogical model, which is the model that underpins the empirical study reported on in this chapter. This figure should be read as taking place over time and not in a single lesson.

The pedagogical sequence depicted in Fig. 7.1 begins with the need to motivate student engagement. Motivation, in this model, owes more to Piaget (1976) than to the scaffolding metaphors of Wood et al. (1976) or Gallimore and Tharp (1990), who require that children are given some kind of lesson hook to engage them in what will follow in the lesson. While this kind of hook can motivate students (listening to a pop song as an introduction to poetry, for example, is a nice way to attract students' attention), we argue here that this is not enough to sustain motivation and is an extrinsic rather than intrinsic motivator. What is required is a form of cognitive conflict, or disequilibrium; that is a dissonance between what the student knows and can do and what that student does not know, motivating the student to act. In this sense, motivation is not a comfortable feeling, but rather one that leads to discomfort and the need to overcome this: through learning from teaching.

Teaching and learning in this model is dialogical in that it is not the teacher alone who holds the space of knowledge; instead knowledge is co-constructed through dialogical interaction where a student is viewed as an active cognising agent who is capable of meaning-making. This unfolds in a context of an ethics of care, where the student is respected and valued and where teacher responses to a student's problem-solving are affirming (Hedegaard & Edwards, 2023). Relational pedagogy in an ethics of care requires an intentional stance towards students, viewing them as active cognising agents, not as passive recipients of knowledge.

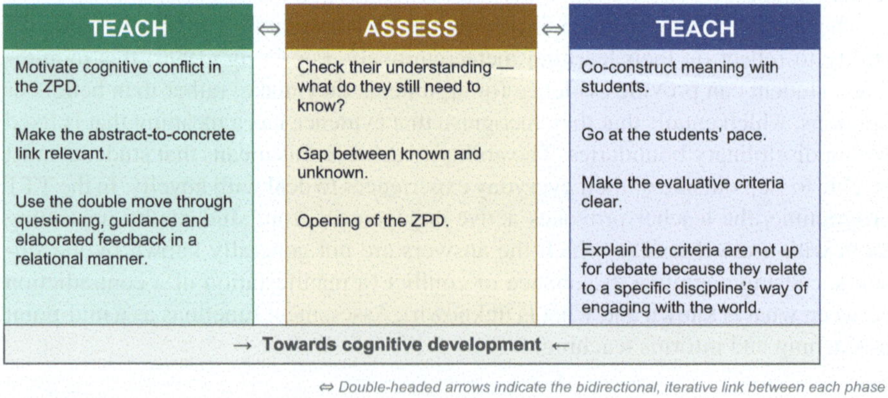

Fig. 7.1 The Teach-Assess-Teach model of developmental pedagogy

This type of teaching aims to develop students through identification of the ZPD, over time. In a teaching episode, a discrete sequence in time in a lesson where concepts are elaborated, a teacher may rely on scaffolding, such as questioning or modelling, to aid the students to engage with the task (Gallimore, 1990). Importantly, scaffolding is not mediation; mediation takes place over time while scaffolding is task specific and can be used in a discrete lesson or on a single task, to help students solve a specific problem. In this model, assessment follows teaching and is aimed at understanding what gaps exist in the students' knowledge base.

The evaluative criteria for assessment, describing what counts as a valid answer, are explicit and elaborated so that a student is aware of where they are going wrong and what they are doing right, and the validity of evidence is ensured by meaning fixed in disciplinary norms. Teaching unfolds at the pace of the students, whose responses to assessments (whether these are formal tests or questions in a lesson) inform teaching episodes. The outcome of teaching in this model is the development of new cognitive structures, not merely the acquisition of content, where 'cognitive structures' refer to neural pathways in the brain (Hardman, 2025).

The study discussed here investigates the efficacy of this model.

The Study of Two Pedagogies in Grade 2 Life Skills: Beginning Knowledge Lessons

In the curriculum Beginning Knowledge is supposed to develop a foundation for future grades science learning. However, there are significant concerns about its ability to do this. One reason for this is the relatively large body of different knowledge that falls under the umbrella of "Beginning Knowledge" such as natural science, social science, and technology, making the epistemological boundaries between these knowledge areas weakly bounded. There is also some doubt about teachers' knowledge base in natural science at the Foundation Phase (FP) impacting on their willingness to teach science which is further compounded by the density of the curriculum that needs to be covered (Dixon et al., 2018).

A case-study approach was used to compare the pedagogical practices of two teachers in the same school. One teacher, Mrs. Jenkins, underwent a 24-hour module in an honours course where she was trained by myself at my university for 2 h a week over 12 weeks to use the pedagogical model discussed in this chapter. The other teacher, Mrs. Smit, did not take this module.

The School

Both teachers reported in this study teach at the same school, which is in an affluent area of Cape Town. The school, which is a co-educational primary school, has approximately 780 students and runs from Grades 1–7. The current study took place in Grade 2 classes. The school is a government school, meaning that it provides the government-approved curriculum and is not considered a private school although parents and students are encouraged to collect extra funds for projects in the school such as setting up a hockey pitch. This school is a quintile 5 school, meaning that it is amongst the most advantaged government schools in South Africa. Each class has approximately 30 students in it. The students all wear a standard school uniform that is prescribed by the school and are not allowed to wear clothing that is not part of the uniform. The school espouses an academic ethos and this is echoed in its motto *per aspera ad astra*, which can be translated as 'through hard work to the stars'.

The Teachers

Mrs. Jenkins is 32 years old and has an undergraduate degree, a postgraduate certificate in education and an honours degree in education. Mrs. Smit is 38 years old and holds an undergraduate degree, as well as a postgraduate certificate in education. Both teachers teach Grade 2, which is in the Foundation Phase. In South African schools, the Foundation Phase encompasses Grades 1–3 and is focused on getting children to read, while Grades 4 onwards focus on reading to learn.

The Students

Grade 2 students are in what Piaget (1976) calls the concrete operational stage of cognitive development, where they rely on concrete instances of objects with which to reason. Children at this age tend to externalise their thinking in problem-solving, giving the researcher a 'window' into their reasoning. The average age in Mrs. Jenkins' class was 8.4 years and the average in Mrs. Smit's class was 8.2 years.

The Subject

I chose to focus my analysis on pedagogy that dealt with science because science underpins many of the crucial skills for navigating the twenty-first century (Rodrigues et al., 2011), and research indicates that children can learn science at a much younger age than was previously thought possible (Fleer, 2011; Morris et al., 2016).

The observations took place in one 'science' lesson taught by each of the two teachers, although there is no designated science subject in the Foundation Phase. Students engage in 2 h of science per week as part of the subject Life Skills, which covers Beginning Knowledge as well as Personal and Social Well-being. According

Table 7.1 Wild and domesticated animals

Wild animals e.g. African wildcat (*Felis lybica*) or lion (*Panthera leo*)	Domesticated animals e.g. house cat (*Felis catus*)[a]
Hunt their own prey	Are dependent on humans
Live in their natural habitat	Live in a human domestic environment
Have not been selectively bred for domestication	Have been selectively bred for domestication

[a]The Latin word '*felis*' means 'cat', while the Latin word '*catus*' means 'domestic'

to the curriculum, 2 h are to be used to cover the topic 'Wild animals' as part of Beginning Knowledge. The specific lessons observed both covered this topic.

Table 7.1 shows three distinct differences between wild and domesticated animals. The two teachers focus on different aspects when they teach the concept of wild animals.

Data Collection and Analysis

Lessons were videotaped and transcribed by a professional service. I viewed the videos alongside the transcriptions to check them for accuracy. I then analysed the linguistic data in the transcriptions. Talk formed the central data for this study as it is in the pedagogical discursive interaction that you can understand what is happening in the lesson (Mercer, 2016).

As I run the university course that Mrs. Jenkins studied, this introduced a potential bias into the research. I therefore also let a colleague code the lessons that I coded to ensure that there was consistency across the coding. Inter-rater reliability was satisfactory at 84%. For ease of reference, Table 7.2 below reproduces Table 3.3 in Chap. 3, making it easier to view coding at a glance.

Table 7.2 illustrates how units of speech, that made sense in themselves, were coded according to various types of language that was of interest to the research.

Results and Discussion

Table 7.3 illustrates the percentage of talk per category coded in the observed lessons. Concepts (8%) than Mrs. Jenkins (17%). Mrs. Smit uses many closed questions (24%) while Mrs. Jenkins also uses closed questions (17%), a technique that allows these teachers to scaffold engagement. Mrs. Smit tends to rely quite heavily on the use of scientific terms (21%) and unelaborated feedback (26%). Mrs. Jenkins, conversely, uses elaborated scientific concepts (16%) and elaborated feedback (22%) in her lessons.

Table 7.2 Analytical codes

Code	Definition	Example
ST	Science term- not elaborated	"Cats are predators"
ESC	Elaborated scientific (abstract) concept	Mrs Jenkins: 'They have retractable claws to hunt very quietly and to protect themselves. And so their bones are different to ours'.
ED	Everyday concept	"My cats eat grass"
CQ	Closed question	"So small house cats eat?"
OQ	Open question	"Can you explain what the word predator means?"
FE	Feedback elaborated	"Mrs. Jenkins. So, they are predators, they are hunters".
FN	Feedback not elaborated.	"Yes, that's right"

Table 7.3 Coding of teacher talk across a lesson

Code	Mrs Jenkins %	Mrs Smit %
Scientific term	8	21
Elaborated scientific concept	16	10
Everyday concept	17	8
Closed question	18	24
Open question	2	–
Feedback elaborated	22	11
Feedback not elaborated	17	26
TOTAL	100%	

While these findings are evidence of both similarities and differences in the two teachers' styles, they do not give a full picture of how the teachers interact with students. Extracts from the transcripts allow a deeper investigation of how talk functions in their lessons.

Extract 7.1: A Challenging Closed Question Leading to Cognitive Conflict

Students are sitting on the mat in front of the teacher. The teacher begins to question the students about what cats are like. She is holding a picture of a cat in her hand and is asking them how a tin of cat food she is also holding relates to the cat. She has a picture of a zebra and a lion on the blackboard. She points to the cat picture in her hand and to the zebra and lion on the blackboard.

1. **Mrs Jenkins**: What do they have in common?
2. Yes, Pearl?
3. **Pearl**: I also want to say they're both animals.
4. **Mrs Jenkins**: OK, cats and that's an animal, but it's not a [*unclear*] thing.
5. **Joe**: They both eat grass.
6. **Kayla**: My cat eats grass.
7. **Mrs Jenkins** Yes, Lila?
8. **Lila:** The zebra eats the grass to make it get bigger.

9. **Mrs Jenkins:** No, that's not what I'm looking for.
10. I'll give you two more tries…
11. **Ryan:** If the cat turns into a lion and it eats that, then that eats the [unclear].
12. **Mrs Jenkins:** OK, now say that again.
13. Say that again to me.
14. So small house cats eat?
15. **Ryan**: Cat food.
16. **Sipho:** When they grow up they eat zebras.
17. **Mrs Jenkins:** Yes, but they don't grow up to be lions.
18. But you mean that lions eat zebras…
19. **Sipho:** And they're the same family.
20. **David**: Huh?
21. **Mrs Jenkins**: Ah, bingo!
22. Lions are part of the same family as a cat.
23. **David:** How come because one is wild?
24. **Jane**: Because they're all different types of cats.
25. You can get wild cats and home cats and trained cats and different cats.
26. **Mrs Jenkins:** Wild cats like lions don't need to be fed by us but our cats at home need us to feed them.
27. **Daniel:** Yes.
28. **Mrs Jenkins:** Please give yourselves a clap.

The teacher begins this extract with an everyday object that children can identify: cat food. As the lesson develops, we can see that her intention is to compare and contrast wild and domesticated animals, in this instance, felines. In lines 12–21 the teacher uses elaborated feedback and closed questions to prompt students towards the correct answer, which she then reiterates in line 22. In lines 12–22 we can also see the teacher using questions and prompts to get at the point she wants to make, which is that cats and lions are part of the same family. This is an interesting exchange in two ways.

Firstly, we can see how Mrs. Jenkins uses closed questions as guides to get to the answer she wants. In this sense, the questions serve as scaffolds towards the right answer. Closed questions can actually open up discussion to the extent that they enable a flow in the lesson between teacher and taught. Classrooms are asymmetrical spaces where power is held by the teacher and not generally shared with students. Where the teacher makes space available for students, even in closed questions, there is a chance for student voices to be heard (Hardman, 2008), and there is a consequent subtle shift in power relations. This is shown in the confidence with which Jane speaks in lines 24 and 25, when she explains the concept to David. It is also shown in David's question in line 20, which indicates that he feels able to speak without fear of being shamed. By asking a question we put ourselves at risk, which takes a level of bravery in a classroom scenario.

Secondly, Mrs. Jenkins is provoking, for at least some students (see David, lines 20–26), a moment of cognitive conflict in that they may not realise that cats and lions form part of the same family. David will now have a notion of felines that are domesticated and felines that are wild. In addition, she has encouraged the students to wrestle with issues of categorisation. Nevertheless, Mrs. Jenkins does not leave students in a state of confusion, as in line 26, she does clarify the concept that she wants students to know, elaborating the scientific concept of wild by stating that these animals do not rely on humans for food.

All this takes place within a safe space where the teacher recognises the contributions of her students. Mrs. Jenkins listens closely and does not hurry her students. She informally assesses the concepts that the students express, and tells them which aspects of their answers are incorrect and which aspects are valuable and should be further refined. This is apparent in lines 16–18 when Sipho says, 'When [cats] grow up they eat zebras.' Mrs. Jenkins shows that part of this statement is incorrect: cats do not grow up to be lions. But she also takes the time to affirm the part that is correct, which is the 'lions eat zebras'. The praise in line 28 is warm and fun, and is an example of contingency management, a form of scaffolding that Gallimore (1990) picks up in his work, where praise or rewards ensure engagement from the students.

Therefore the exchange in Extract 7.1 illustrates how a teacher successfully harnesses an everyday concept that the students are familiar with (cat food is meat that domestic cats eat) and intertwines this with the abstract scientific concept she is teaching (wild animals hunt for their meat and do not live with humans).David's question is clearly a good question in that it shows that he is grappling with knowledge that is outside of merely understanding what a wild animal is (which is the outcome of the curriculum task) and is able to conceptualise of different understandings of family. What he appears to be engaged with, then, is not what is narrowly prescribed but it is something more, something beyond the curriculum requirements.

Extract 7.2 is from a lesson by the teacher who was not trained in the teaching method outlined in this chapter, Mrs. Smit.

Extract 7.2: Easy Closed Questions and the Introduction of Scientific Terminology

Mrs Smit has spent 6 min settling the children on the mat. She now begins her lesson. This is a lesson about wild animals. She holds up a picture of a lion.

1. **Mrs Smit:** Russel, what animal is this?
2. **Students:** A lion.
3. **Mrs Smit:** A lion.
4. Good.
5. And what noise does a lion make? *Students roar.*
6. **Mrs Smit:** And where do we find lions?
7. **Students:** On land.
8. **Mrs Smit:** On land, good boy.
9. Good boys and girls.
10. We find lions on land.
11. And do lions like water?
12. **Students:** No.

13. **Mrs Smit:** But do they drink water?
14. **Students:** Yes.
15. **Mrs Smit:** Yes.
16. And what family is the lion from?
17. **Students:** Cats.
18. **Mrs Smit:** The cat family.
19. Good, so, so, so good.
20. The lion is from the…
21. **Students:** Cat family.
22. **Mrs Smit:** Cat family.
23. Right.
24. Now what can you tell me?
25. What do lions eat?
26. **Students:** Meat.
27. **Mrs Smit:** And how do they get the meat?
28. Do they go to Spur?
29. **Students:** No.
30. **Mrs Smit:** Do they go to Panarottis for their meat?
31. **Students:** No.
32. **Mrs Smit:** Do they eat people?
33. Oh no.
34. Shhtt. Shhtt. *Children are shouting out answers.*
35. I said, ‘No screaming’.
36. I said, ‘No screaming out’.
37. Zip your lip, Tamsyn!
38. Yes, Surina?
39. **Surina:** The lion went and then when the animals can’t see them, then they bite the animals.
40. **Mrs Smit:** Yes, good girl.
41. Surina is describing what we call hunting.
42. Everybody say, ‘hunting’.
43. **Students:** Hunting.
44. **Mrs Smit:** Now hunting is what lions do when they want to catch prey.
45. And prey is another word for food.
46. You have a predator which is a lion.
47. Everybody say ‘predator’.
48. **Students:** Predator.
49. **Mrs Smit:** Yes, and then you have the prey.
50. Everybody say ‘prey’.
51. **Students:** Prey.
52. **Mrs Smit:** And it’s not this kind of pray. *Holds her hands together and closes her eyes imitating praying.*
53. OK?
54. It’s not this kind of pray; it is food.

In lines 1–8 the teacher uses closed questions to ascertain what students know about lions. The evaluation in these lines is unelaborated with the teacher either restating the answer given by the students or simply acknowledging it as correct. The students appear to know that the lion is part of the cat family (line 17), and by line 18 the teacher has made this clear.

In line 28 the teacher tries to recruit the students' everyday concept of eating out at restaurants such as the Spur or Panarottis.[3] The children are familiar with these restaurants, but do not appear aware of what Mrs. Smit requires other than a 'no' until Surina, in line 39, indicates that lions hunt for their food ('bite the animals'). In lines 41 and 42 the teacher elaborates the answer given by the student so that the students all know the word 'hunting'. Continuing in this vein, in lines 44–51 the teacher elaborates the scientific concepts of prey and predator by harnessing the concept of hunting. Note in this teacher's lesson on wild animals that she doesn't seek to compare and contrast them to domesticated animals but seems rather to want to define them in terms of hunting.

She uses a traditional question-and-response technique when she gets students to chorus the words 'hunting', 'predator' and 'prey'. It is unclear if the students understand these concepts or are merely able to repeat the scientific terms. The teacher then indicates that 'prey', which wild animals hunt, is not the same as its homophone 'pray', which is a religious term. While she may be introducing an element of humour here, her instruction to Tamsyn to 'zip' her lip in lines 37 is likely to curtail much student engagement.

This extract is characteristic of Mrs. Smit's pedagogy across the observed lessons. She begins her lessons with an attempt to scaffold engagement by presenting students with something that is of interest, such as the picture of a lion in this lesson. She controls the pace of the lesson, and her feedback is generally unelaborated. There is no evidence of cognitive conflict in the extract. Instead Mrs. Smit is determined to increase the students' scientific vocabulary. She uses informal assessment to check that students know the terms she has chosen to emphasise.

While Mrs. Smit had not been trained to use a the TAT model aimed towards an inclusive developmental pedagogy, Mrs. Jenkins had. Ms. Jenkins made use of many of the techniques she was taught in the course she studied on this pedagogical model. She created cognitive conflict and used scaffolds for task-specific problems. She recruited everyday concepts and elaborated scientific concepts, and encouraged students to view these two as dialectically entailed. In her class children felt able to pose questions.

Conversely, Mrs. Smit tends to control pace in her lesson and offers no chance (across four observed lessons) for students to pose their own questions. The class is a traditional one where there are a lot of closed questions and choral repetition of responses. There does not appear to be much space for the inclusion of children's unique voices.

[3] These are popular restaurants in South Africa.

This in no way suggests that Mrs. Smit is not a good teacher. Her content knowledge is sound, and she tries to keep children engaged in her lessons through closed questions. She covers the content required by the curriculum at an appropriate pace, and does not allow students to derail the lesson. In the above extracts, the students probably acquire more scientific terms in Mrs. Smit's lesson than Mrs. Jenkins's lesson, however, scientific terms alone are hollow and meaningless. Without meaning, remembering a concept requires that one rote memorise, rather than understand the concept. In most cases the students are not co-constructing meaning, which means they are acquiring mere definitions of concepts rather than 'owning' these definitions (as discussed in Chap. 1).

In contrast, while Mrs. Jenkin's lesson was less tailored to the curriculum, and the pace was slower and fewer terms were learnt, the learners in her lesson were required to think harder, to identify patterns beyond what was required in the curriculum, and to phrase them in their own words. Thus they were encouraged to engage in more advanced cognitive work and develop more advanced neural pathways. Given the evidence we have for the value of an active inclusive pedagogy (Hedegaard & Edwards, 2023; Hoadley 2017), which allows for a weakening of pace and power relations as well as an increase in dialogue, the model of pedagogy applied by Mrs. Jenkins is more suggestive of positive long-term learning outcomes than that applied by Mrs. Smit.

Conclusion

This chapter sets out an argument for a model of dialectical pedagogy that develops students' understanding in class. While much of traditional schooling is aimed at reinforcing content knowledge, this model advocates a pedagogy that develops students both cognitively and emotionally by situating pedagogy in a context of care. The model draws on the work of Vygotsky (2012) and Hedegaard's (1998) double move and uses assessment as a teaching mechanism rather than as an end point of learning.

Empirical data from two teachers in the same school, teaching the same subject, were presented. One of the teachers in this chapter, Mrs. Jenkins, had been taught to use the TAT model and the other had not. This model is aimed at inclusive pedagogy where learning leads development through a cycle of teaching/testing/teaching. Based in dialectical logic this model sees teaching and learning as single sides of one coin, inextricably linked and producing contradictions as novel knowledge is developed predominantly in discussion. Findings indicate that the use of this model can provoke cognitive conflict, harness the use of abstraction and everyday experiences for understanding, and is situated in a context of an ethics of care.

This is a small case study and there is more work to be done to ascertain the impact of this kind of pedagogy on a wider audience. However, these early findings provide some evidence that this pedagogy can create a safe space, give more weight to students' voices and give them practice in knowledge production that will prepare them better for later stages of schooling and beyond.

References

Ashby-King, D. T., & Anderson, L. B. (2022). "It gives you a better chance of getting a good job": Memorable messages, anticipatory socialization, and first-year college students' understandings of the purpose of college. *Communication Education, 71*(1), 2–20.

Craig, A. P. (1996). Education for all. *South African Journal of Higher Education, 10*(2), 47–55.

Department of Basic Education. (2024). *South African 2023 trends in international mathematics and science study (TIMSS) highlights report.* Department of Basic Education. https://www.education.gov.za/Portals/0/Documents/Reports/2024/TIMSS%202024.pdf

Dixon, K., Janks, H., Granville, S., Kell, C., Makalela, L., & Tyler, R. (2018). A critical analysis of CAPS for life skills in the foundation phase (Grades R-3). *Journal of Education, 71*, 1–20. https://doi.org/10.17159/2520-9868/i71a01

Fleer, M. (2011). "Conceptual play": Foregrounding imagination and cognition during concept formation in early years education. *Contemporary Issues in Early Childhood, 12*(3), 224–240.

Gallimore, R., & Tharp, R. (1990). Teaching mind in society: Teaching, schooling. *Vygotsky and education: Instructional implications and applications of sociohistorical psychology, 175.* Cambridge University Press.

Hardman, J. (2008). Researching pedagogy: An activity theory approach. *Journal of Education, 45*(1), 65–95.

Hedegaard, M. (1998). Situated learning and cognition: Theoretical learning and cognition. *Mind, Culture, and Activity, 5*(2), 114–126.

Hedegaard, M., & Edwards, A. (2023). *Taking children and young people seriously: A caring relational approach to education.* Cambridge University Press. https://doi.org/10.1017/9781108980869

Hoadley, U. (2017). *Pedagogy in poverty: Lessons from twenty years of curriculum reform in South Africa.* Routledge.

Kanjee, A., Chetty, M., Shozi, N., Ahmed, A., & Correa, F. (2025). *TIMSS 2023: Highlights report South Africa.* International Association for the Evaluation of Educational Achievement. https://www.iea.nl/publications/study-reports/national-reports-iea-studies/timss-2023-highlights-report-south-africa

Mercer, N. (2016). Education and the social brain: Linking language, thinking, teaching and learning. *Éducation et didactique, 10*(10-2), 9–23.

Morris, A., Hardman, J., & Jacklin, H. (2016). School science for six-year-olds: A neo-Vygotskian approach to curriculum analysis. *Journal of Education, 64*, 1–26.

Piaget, J. (1976). Piaget's theory. In *Piaget and his school: A reader in developmental psychology* (pp. 11–23). Springer Berlin Heidelberg.

Rodrigues, A., Camillo, J., & Mattos, C. (2011). Cultural-historical activity theory and science education: Foundational principals and potentialities. *Scientific Papers, 778*, 191–200.

Vygotsky, L. S. (1978). *Mind in society: The development of higher psychological processes* (M. Cole, V. John-Steiner, S. Scribner, & E. Souberman, Eds.). Harvard University Press.

Vygotsky, L. S. (2012). *Thought and language* (Vol. 29). MIT Press.

Wood, D., Bruner, J. S., & Ross, G. (1976). The role of tutoring in problem solving. *Journal of Child Psychology and Psychiatry, 17*(2), 89–100.

Chapter 8
Managing the Potential Within Contradictions to Guard Against Regressive Outcomes

Introduction

As CHAT researchers, our central argument in this book has been, unsurprisingly, that contradictions, far from being obstacles to overcome or problems to eliminate, constitute a mechanism of development and transformation within human activity systems. Through detailed theoretical exposition, methodological guidance and rich empirical cases drawn from educational contexts in South Africa, we have demonstrated how contradictions serve as sites of dynamic change that researchers and practitioners can identify, analyse and potentially leverage for progressive transformation. However, in this chapter we add the caveat that, while progressive transformation is at the heart of CHAT, there is always a very real threat of regression, where activity systems change to resemble prior, less empowering versions.

A Dialectical Perspective on the Power of Contradictions in Educational Contexts

As discussed in Chap. 2, Hegel perceived a dialectical relationship to exist between a concept and its opposite concept, and that their conflict could be overcome by acknowledging what was valid in both and clarifying their interrelationship (Hegel, 1977; Singer, 2001). This thesis, antithesis and synthesis format is a useful way of framing what we have already said about the power of contradictions before we discuss key insights from our research and the implications of these for educational policy and practice.

J. Hardman, E. Rzyankina, *Contradictions in Cultural-Historical Activity Theory*, https://doi.org/10.1007/978-3-032-28650-5_8

CHAT'S Thesis: Contradictions Can Drive Progressive Change

In Part 1, we traced the historical evolution of contradiction as a philosophical and analytical concept, revealing that contradictions have always been more than logical inconsistencies or rhetorical devices. From Heraclitus's recognition that strife generates change, through Hegel's dialectical method where concepts contain their own negation, to Marx's materialisation of dialectics in analysing capitalism's inherent tensions, contradictions have been understood as generative forces that propel systems forward.

Vygotsky used a dialectical approach to emphasise that children develop through the conflict between their actual and potential capacities. His zone of proximal development (ZPD) implicitly acknowledges contradiction as the space where what children can do independently clashes in a productive way with what they might achieve with assistance. Leontiev's elaboration of activity theory identified tensions within human activities that were useful to society (such as hunting for food). However, it was Engeström's idea of learning as 'expansive' (1987, 2005) that epitomises CHAT's optimism towards the possibilities inherent in contradictions.

We have endorsed the usefulness of Engeström and Sannino's (2011) discursive manifestations of contradictions framework, which shows how contradictions become visible in participants' language and actions. We have also supported the view that when contradictions are resolved in a progressive way they change the activity into a more historically mature[1] activity. It is clear that contradictions are potentially transformative and can lead not only to change but to progress.

The Logical Antithesis: Contradictions Can Drive Regressive Change

'Contradiction' is not a word that has positive connotations; in fact, it has negative connotations. This is partly because, from an Aristotelian perspective, a contradiction is a logical flaw. However, it is also because a contradiction leads to tension, as CHAT theorists recognise, and tensions are not always constructively resolved, as news websites and history books often show. The amount of effort that is put into the constructive resolution of tensions is a sign of the fear of what will happen if these tensions lead to negative outcomes.

Engeström and Sannino (2011) name four ways that contradictions can make themselves apparent in what people say and do: 'dilemmas', 'conflicts', 'critical conflicts' and 'double binds'. These are all terms with negative connotations, because they show that these people are frustrated by their situations. People in such situations want to change them, but they will not always choose to do so in a way

[1] We differentiate here between chronological change and socially beneficial change, where historically mature refers to socially beneficial change.

that is beneficial to society as a whole, and may choose ways that are not even in their own self-interest. Contradictions increase volatility, and tense situations may erupt in unpredictable ways.

In South African history, some Black people in the Cape had the right to vote before their names were removed from the common rolls in 1936. They only regained the right to vote in this region in 1994. This illustrates how the tensions caused by contradictions have the potential to cause activity systems to regress as well as progress. In an educational context, confronting contradictions can also lead systems to resemble prior, more restrictive forms of practice (Hardman, 2022).

The Synthesis: Contradictions Can Drive Progressive or Regressive Change

The antithesis complicates any assumption that contradictions are inherently progressive forces and shows it to be simplistic. To synthesise the thesis and antithesis we must acknowledge that contradictions can drive both progressive and regressive change.[2] This means that researchers, policymakers and practitioners with a transformative agenda must attend carefully to how contradictions are managed within specific contexts.

Empirical Insights from our CHAT Research on STEM Education in South Africa

Examining the lived experience of contradictions shows that they are not merely analytical constructs imposed by researchers but real, consequential features of social practice that participants themselves recognise and struggle with. Insights from educational contexts can help us learn how to identify, release and harness the potential of contradictions so that these contradictions power progress rather than the opposite. We hope that the empirical chapters in Part 2 help to achieve this, and here we revisit the main findings of our research on activity systems in a university of technology and in two primary schools, and we then interpret what these insights indicate once they are viewed together.

[2] The thesis, antithesis and conclusion form a syllogism, a logical structure discussed in Chap. 2:

Premise 1: Contradictions can drive progressive change.
Premise 2: Contradictions can drive regressive change.
Conclusion: Contradictions can drive progressive or regressive change.

Empirical Insights on Activity Systems in a University of Technology

The introduction of e-textbooks in engineering education at a South African university of technology triggered contradictions at multiple levels as this novel mediational artefact was integrated into existing pedagogical practices (see Chaps. 4 and 5).

The primary contradiction, between learning Physics with e-textbooks versus learning how to operate e-textbooks, exemplified tool-object reversal, where the mediational artefact itself becomes the focus of activity rather than serving as a transparent resource for achieving the intended object. This contradiction speaks to a fundamental challenge in educational technology: how can digital tools enhance learning when significant cognitive and practical resources must first be devoted to mastering the technology itself?

Secondary contradictions also emerged between various elements of the students' activity system, such as conflicts over access arrangements, which illustrated how introducing a novel tool has repercussions throughout an activity system, creating new tensions between existing practices and emerging possibilities.

Particularly striking was the tertiary contradiction between students' established digital literacy practices (for example, relying on Google) and the new digital literacy practices required by e-textbooks. This contradiction reveals how educational innovations do not enter neutral spaces but rather encounter deeply sedimented habits, skills and preferences shaped by students' entire educational and cultural histories. The resolution of such contradictions cannot be achieved through a brief training session but requires sustained transformation of practice over time.

The quaternary contradictions between students' and lecturers' activity systems highlighted how the same tool, the e-textbook, was understood differently by different participants. While students struggled with operational competencies, access issues and new digital literacy practices, lecturers viewed e-textbooks as resources for drill-and-practice or content delivery. This misalignment between neighbouring activity systems sharing overlapping objects demonstrates why educational change is so difficult: different stakeholders bring different motives, tools, rules and divisions of labour to supposedly 'shared' activities.

Empirical Insights on Teaching in Primary Schools

Contradictions emerged in both studies on pedagogy in primary schools (see Chaps. 6 and 7). These looked at what happens when computer technology is introduced into a Grade 6 Mathematics classroom and how the topic of 'Wild animals' is taught to Grade 2 students in Life Skills.

In the study of the teaching of Mathematics there was a striking disjuncture between the teacher's intended pedagogy, which was characterised by constructivist principles and collaborative learning, and his enacted pedagogy in computer-based lessons, which often reverted to highly directive, teacher-controlled instruction focused more on managing behaviour and teaching technical skills than on developing mathematical understanding. This reversion to traditional, restrictive pedagogical forms in the presence of novel technology represents a crucial finding that complicates optimistic narratives about technology as inherently transformative. Multiple contradictions contributed to this regressive change: primary contradictions within the object of lessons (mathematical understanding versus technical skill development), secondary contradictions between teachers' espoused pedagogical beliefs and the rules constraining computer-based instruction, and tertiary contradictions between traditional classroom pedagogy and the new practices potentially enabled by technology.

In the study of the teaching of the lessons on wild animals, two types of pedagogy were compared: teaching using the Teach-Assess-Teach (TAT) model, which relies on contradictions in the form of cognitive conflict that manifests in dialogue, and teaching using the traditional approach in South African schools, which emphasises the acquisition of key terms on a particular topic. The study examined the differences in classroom pedagogy between a qualified teacher who had been taught to use the TAT model with a qualified teacher who had not been taught to use this particular model, but who had years of experience using a more traditional approach. The comparison showed that the teacher using the TAT model caused her students to think more about how we categorise, while the teacher focusing on terminology caused her students to learn more terms. Of course, it is important to learn scientific terms as they give ostensive definition to a concept, however, without interpenetration of the everyday concept, they are essentially meaningless, hollow terms.

In Chap. 7, Ms. Smits' reliance on traditional pedagogy is not dissimilar to Mr. B, in Chap. 6, who reverts to traditional methods of teaching due to the time constraints imposed by having to cover a large, dense curriculum in a relatively short amount of time. The reversion to traditional behaviourist type teaching then, is not a feature of the teacher, per se, but rather, a feature of the activity they are engaged in.

Synthesising the Insights from Both Sets of Research

The case studies reported in the book, ranging from University students to primary school students, illustrate a common theme; training is needed when novel technology is introduced into an activity. While the university students did not feel they were given the requisite training for the technology used, the school students were trained to use the technology, at the expense of learning mathematics. What we see then is that the introduction of novel technology into an activity is impacted by various nodes with the activity.

This highlights the interconnected nature of contradictions, which form complex, nested patterns where contradictions at one level both produce and are produced by contradictions at other levels. Effective intervention thus cannot address single contradictions in isolation but must engage with these complex systemic patterns.

All the studies illustrate the importance of historical analysis in understanding contradictions. Every contradiction identified in our empirical chapters had deep historical roots; in apartheid education policies, in the global political economy of educational publishing, in long-established pedagogical traditions, and in students' prior educational experiences. Therefore every 'historically accumulating structural tension', to use Engeström's phrase, requires not only immediate practical responses but also engagement with the historical forces that produced them.

Implications for Research, Policy and Practice

Our findings regarding regressive as well as progressive responses to contradictions complicate developmental narratives within CHAT. While Engeström's theory of expansive learning emphasises how contradictions can catalyse qualitative transformations towards more expansive forms of activity, our empirical cases demonstrate that contradictions can equally generate defensive reactions, reversion to familiar practices or intensification of control. This recognition demands more nuanced theorising about the conditions under which contradictions lead to an expansion of learning versus a contraction of learning.

Nevertheless, our research does have practical implications for policymakers, educational leaders and teachers seeking to leverage technology and pedagogical innovation for improved educational outcomes. The contradictions identified in our empirical chapters point towards specific areas requiring attention and intervention.

Firstly, the introduction of novel educational technologies must be accompanied by sustained, substantive professional development that goes beyond technical training. Teachers need opportunities to develop not just operational competencies but also pedagogical reimagination; understanding how new tools might enable different forms of teaching and learning while being supported in navigating the inevitable contradictions that emerge. (A potential model of such a pedagogy is offered in Chap. 7.)

Secondly, educational technology initiatives must attend seriously to questions of access, infrastructure and equity. As our findings demonstrated, providing devices or digital resources without ensuring reliable connectivity, adequate devices for all students and culturally appropriate content may intensify rather than resolve educational inequalities. Contradictions around access are not peripheral technical issues but central challenges requiring systematic institutional responses.

Thirdly, curriculum and assessment policies must be designed with an awareness of how they shape the object of educational activity. When rules around curriculum coverage, pacing and standardised assessment create secondary contradictions with teachers' understanding of effective pedagogy or students' actual learning needs, regressive responses become likely. Creating space for teachers and students to pursue depth over breadth, understanding over memorisation, and authentic engagement over compliance requires rethinking systemic rules and evaluation criteria.

Finally, educational change initiatives should anticipate and plan for contradictions rather than treating them as unexpected obstacles. The Developmental Work Research and Change Laboratory methodologies, as well as other CHAT-informed intervention approaches, explicitly use contradiction analysis as a foundation for collaborative transformation. By helping participants identify and collectively analyse contradictions in their own activity systems, these approaches can support more sustainable, locally appropriate innovations than top-down implementation models.

Issues for Further Investigation

While this book has contributed to both theoretical understanding and practical application of contradiction analysis in educational research, it inevitably leaves questions unanswered and new questions opened.

In relation to methodology, the challenge of identifying contradictions and distinguishing them from related phenomena remains complex. While analytical frameworks exist, contradiction analysis ultimately requires interpretive judgement informed by deep contextual knowledge and theoretical sensitivity. Further methodological development, including more explicit criteria for warranting claims about contradictions and more systematic approaches to analysing how contradictions resolve or transform over time, would strengthen this analytical approach.

In relation to theory, CHAT's analytical focus on systemic contradictions could be enriched by greater attention to the subjective dimensions of navigating contradictions. In addition, the relationship between contradictions and transformation remains incompletely theorised. We will highlight three key questions in this regard:

- Under what specific conditions do contradictions catalyse expansive learning and progressive change rather than defensive responses and regressive reversion?
- What forms of intervention, support or collective action enable productive engagement with contradictions?
- How long do transformations take, and what institutional conditions sustain them beyond initial enthusiasm?

How the Global South Can Use the Concept of Contradiction Fruitfully for Educational Progress

Questions of social justice emerged as central concerns throughout the empirical chapters. The South African context, marked by profound historical inequalities in educational opportunity and resource distribution, shaped how contradictions manifested and how they might be addressed. The digital divide, unequal access to devices, internet connectivity and digital literacy created and intensified contradictions that would not exist in more privileged contexts, or in less privileged contexts.

Cultural and linguistic contradictions also surfaced, particularly regarding e-textbook content drawn primarily from American and European contexts. Lecturers identified the disconnect between these examples and students' lived experiences as a significant barrier to learning, pointing towards the need for culturally responsive educational resources that reflect students' own languages, contexts and knowledge systems.

These findings align with broader calls within CHAT scholarship, particularly from scholars like Stetsenko (2021), for theories that engage seriously with questions of power, privilege and injustice rather than treating them as peripheral concerns. Contradictions in educational activity systems in contexts such as South Africa are not merely technical or pedagogical issues, but are deeply entangled with colonial histories, economic inequalities and ongoing struggles for epistemic justice.

The concept of contradictions, understood through a Global South lens,[3] thus becomes a powerful analytical tool for revealing how systemic inequalities are embedded within and reproduced through educational practices. Primary contradictions between use value and exchange value take on particular significance in contexts where educational resources are often commodified and marketed in ways that exacerbate rather than ameliorate inequality. Secondary contradictions between students or teachers as subjects and the rules of educational systems reveal how curriculum policies, assessment regimes and institutional structures may conflict with the actual learning needs and experiences of marginalised students.

Perezhivanie *and the Dramatic Frame of Personality as Tools for the Global South*

While Vygotsky is associated by many in education with only one concept, the ZPD (see Chap. 1), his concepts of *perezhivanie* and the dramatic frame of personality (see Chap. 2) are receiving increasing scholarly attention. This is particularly the case in the Global South as researchers seek conceptual tools that can account for the lived realities of students and communities in contexts marked by historical, cultural and socio-economic complexity. Vygotsky conceived of *perezhivanie* to capture the indivisible unity between environmental conditions and the subjective experience of the individual, emphasising that development is not reducible to external influences alone, nor to internal psychological states treated in isolation. In parallel, the notion that personality is formed within a dramatic frame draws

[3] The distinction between the Global North and the Global South is easy to criticise as simplistic because of the wealth disparities within many of Global South countries, but a Global South lens is nevertheless a powerful tool that acknowledges how the wealthier nodes in Global South countries are economically linked to their poorer regions.

attention to the collisions, tensions and negotiations that characterise human conduct, highlighting that development unfolds through crises, turning points and situations of struggle.

When psychological development is examined through the lens of contradiction, the link between *perezhivanie* and the dramatic frame of personality becomes apparent. Contradictions manifest not simply as epistemological puzzles or logical inconsistencies, but as lived tensions that provoke transformation: the individual encounters obstacles or disequilibria that cannot be resolved through existing modes of action, and so it is within a moment of crisis that new forms of conduct and meaning emerge. *Perezhivanie* is the experiential unit in which this process becomes psychologically real; the dramatic frame of personality is the broader structural context in which contradictions unfold across time and within social relations. Together, they offer a robust vocabulary for explaining how development is propelled by internal tensions rather than by harmonious adaptation.

For scholars in the Global South, these concepts resonate deeply because the landscapes of learning, identity and social participation are saturated with contradictions arising from colonial legacies, cultural hybridity, socio-economic inequality and structural transformation. In such settings, the developmental trajectory of learners cannot be understood purely through universalist models that abstract away from lived experience. Instead, concepts such as *perezhivanie* and the dramatic frame foreground how historical, cultural and institutional forces are refracted through personal experience, producing a situated developmental drama in which crises, breakthroughs and contradictions are constitutive rather than incidental. *Perezhivanie* and the dramatic frame of personality assert the significance of conflict in the formation of mind, identity and agency.

Looking forward, the integration of *perezhivanie*, the dramatic frame of personality and the dialectical analysis of contradictions offer a promising research agenda for educational psychology, developmental theory and pedagogy in the Global South and beyond. These concepts challenge reductionist views of learning as the acquisition of discrete competencies and instead advocate for a science of development attentive to conflict, meaning, value and lived experience. They also open the possibility of methodological innovation: analysing contradictions through discursive manifestations, tracing developmental crises through narrative methods, and examining *perezhivanie* in real situations through ethnographic and interventionist approaches.

Detailed attention to how contradictions manifest in diverse contexts enrich understanding of these phenomena beyond abstract categorisation. We believe that the future of CHAT lies not in the stabilisation of its existing concepts, but in their continued elaboration and renewal in response to contemporary educational, cultural and political challenges. The Global South has an essential role to play in this process, precisely because it brings into view developmental dramas and contradictions that existing frameworks have often marginalised. By placing *perezhivanie* and the dramatic frame of personality at the centre of developmental theory, we move closer to a psychology capable of addressing real human problems in their full historical, cultural and experiential richness.

Conclusion

This book has argued that contradictions, understood as historically accumulated structural tensions within and between human activity systems, constitute essential features of social practice that can serve as mechanisms of development and change. Through conceptual clarification, methodological guidance and empirical illustration, we have demonstrated how contradiction analysis provides powerful tools for understanding why educational progress is so difficult and how it might be more effectively pursued.

The cases examined in this book revealed that introducing novel tools into educational activity systems inevitably generates contradictions that reverberate throughout these systems. These contradictions are not failures of implementation but unavoidable features of transformation. How they are understood, engaged with and potentially resolved determines whether change leads towards expansion or contraction, inclusion or exclusion, justice or inequality.

For researchers, policymakers, teachers and other practitioners working towards more equitable, effective and humane educational systems, attention to contradictions offers a path forward that neither naively celebrates innovation nor cynically dismisses possibility. It requires recognising that transformation is difficult, that tensions are inevitable, that regressive responses are understandable, and yet that progressive change remains possible when contradictions are collectively identified, critically analysed and thoughtfully addressed. This is the promise and challenge that CHAT's focus on contradictions offers to those committed to educational transformation in an unequal world.

References

Engeström, Y. (1987). *Learning by expanding: An activity theoretical approach to developmental research*. Orienta-Konsultit.

Engeström, Y. (2005). *Developmental work research: Expanding activity theory in practice*. Lehmanns Media.

Engeström, Y., & Sannino, A. (2011). Discursive manifestations of contradictions in organizational change efforts: A methodological framework. *Journal of Organizational Change Management, 24*(3), 368–387. https://doi.org/10.1108/09534811111132758

Hardman, J. (2022). A cultural-historical pedagogical model as a potential developmental tool in schools. *European Journal of Education and Pedagogy, 3*(3), 191–198.

Hegel, G. W. F. (1977). *Phenomenology of spirit* (A. V. Miller, Trans.). Oxford University Press. (Original work published 1807).

Singer, P. (2001). *Hegel: A very short introduction*. Oxford University Press. (Original work published 1983).

Stetsenko, A. (2021). Scholarship in the context of a historic socioeconomic and political turmoil: Reassessing and taking stock of CHAT. Commentary on Y. Engeström and A. Sannino "from mediated actions to heterogenous coalitions: Four generations of activity-theoretical studies of work and learning". *Mind, Culture, and Activity, 28*(1), 32–43. https://doi.org/10.1080/10749039.2021.1874419

Appendices

Appendix A: CHAT Example of Codebook for Deductive Coding

Analytical concept/ unit of analysis	Characteristics for analytical concept	Component, parts	Description	Code	Explanation of what to look for in the data
Subject (Yamagata-Lynch, 2010; Uden, 2007)	Subject – respondent		The subject is individual or groups of individuals involved in activity. 1. Engineering students; 2. Engineering lecturers at university of technology.	Subject	Who are the people participating in the activity system?
Object (Yamagata-Lynch, 2010; Uden, 2007)	Conducted to move the practice. Object – stimulus	Object/ motive – activity Goal – action Conditions – operation	The object is 'the true motive'. The object characterises sub-activities in the activity system and are eventually converted into outcomes.	Object	What is the goal or motive of the digital reading activity?

J. Hardman, E. Rzyankina, *Contradictions in Cultural-Historical Activity Theory*, https://doi.org/10.1007/978-3-032-28650-5

Appendix B: Example of Codebook for Contradiction Concept

Analytical concept/ unit of analysis	Component, parts	Characteristics of analytical concept	Description of components	Code	Explanation of what you are you going to look for in the data (transcript)
Contradictions of activity (the fourth principle) (Engeström, 2001; Engeström & Sannino, 2010)	Manifestations of contradictions: **dilemma, conflict critical conflict, double bind**	Contradictions are conflicts and tensions within and between activity systems, and sources of change and development. Contradictions form the basis for improved forms of activity, by overcoming them new ways of doing things are developed. Contradictions are conventionally classified as primary, secondary, tertiary or quaternary, as defined in the main text.	**Primary contradictions** exist in capitalism because of the use value and exchange value of commodities. They are observed within the nodes of an activity (the subject, object, tools, community, rules or division of labour). **Secondary contradictions** appear when a system adopts a new element from outside. These are observed between the nodes of an activity. **Tertiary contradictions** result from the effect of a culturally more advanced activity introducing an alternative outcome that supersedes that of the central activity. **Quaternary contradictions** are observed among differing activity systems where a component of the main activity is in conflict with the components of other activities.	Contr MoCs: Dilemma; Conflict; Critical conflict; Double bind	Primary cycle linguistic cues Secondary cycle

References

Al-Abdullatif, A. M., & Gameil, A. A. (2021). The effect of digital technology integration on students' academic performance through project-based learning in an e-learning environment. *International Journal of Emerging Technologies in Learning, 16*(11), 189–210. https://doi.org/10.3991/ijet.v16i11.19421

Ardila, A. (2018). Origins of executive functions. In A. Ardila (Ed.), *Historical development of human cognition: A cultural-historical neuropsychological perspective* (pp. 107–134). Springer. https://doi.org/10.1007/978-981-10-6887-4

Ashby-King, D. T., & Anderson, L. B. (2022). "It gives you a better chance of getting a good job": Memorable messages, anticipatory socialization, and first-year college students' understandings of the purpose of college. *Communication Education, 71*(1), 2–20. https://doi.org/10.1080/03634523.2021.1952462

Bakhtin, M. M. (1984). *Problems of Dostoevsky's poetics*. University of Minnesota Press.

Barton, D., & Hamilton, M. (1998). *Local literacies: Reading and writing in one community*. Routledge.

Bateson, G. (1972a). *Steps to an ecology of mind: Collected essays in anthropology, psychiatry, evolution, and epistemology*. Jason Aronson.

Bateson, G. (1972b). The logical categories of learning and communication. *Steps to an Ecology of Mind*, 279–308.

Birgin, O., & Topuz, F. (2021). Effect of the GeoGebra software-supported collaborative learning environment on seventh-grade students' geometry achievement, retention and attitudes. *Journal of Education in Science, Environment and Health, 7*(4), 282–299.

Boadu, S. K., & Boateng, F. O. (2024). Enhancing students' achievement in mathematics education in the 21st century through technology integration, collaborative learning, and student motivation: The mediating role of student interest. *Eurasia Journal of Mathematics, Science and Technology Education, 20*(11), em2534. https://doi.org/10.29333/ejmste/15622

Bright, A., Welcome, N. B., & Arthur, Y. D. (2024). The effect of using technology in teaching and learning mathematics on student's mathematics performance: The mediation effect of students' mathematics interest. *Journal of Mathematics and Science Teacher, 4*(2), em059. https://doi.org/10.29333/mathsciteacher/14309

Britain, G. (2001). *ImpacT2: Emerging findings from the evaluation of the impact of information and communication technologies on pupil attainment.*

Campbell, S., Greenwood, M., Prior, S., Shearer, T., Walkem, K., Young, S., Bywaters, D., & Walker, K. (2020). Purposive sampling: Complex or simple? Research case examples. *Journal of Research in Nursing, 25*(8), 652–661. https://doi.org/10.1177/1744987120927206

J. Hardman, E. Rzyankina, *Contradictions in Cultural-Historical Activity Theory*, https://doi.org/10.1007/978-3-032-28650-5

Campuzano, L., Dynarski, M., Agodini, R., & Rall, K. (2009). *Effectiveness of reading and mathematics software products: Findings from two student cohorts*. NCEE 2009-4041. National Center for Education Evaluation and Regional Assistance.

Case, J. M., & Light, G. (2011). Emerging research methodologies in engineering education research. *Journal of Engineering Education, 100*(1), 186–210. https://doi.org/10.1002/j.2168-9830.2011.tb00008.x

Chauhan, S. (2017). A meta-analysis of the impact of technology on learning effectiveness of elementary students. *Computers & Education, 105*, 14–30.

Cheung, A. C., & Slavin, R. E. (2013). The effectiveness of educational technology applications for enhancing mathematics achievement in K-12 classrooms: A meta-analysis. *Educational Research Review, 9*, 88–113.

Christie, P., & Collins, C. (1982). Bantu education: Apartheid ideology or labour reproduction? *Comparative Education, 18*(1), 59–75. https://doi.org/10.1080/0305006820180106

Cohen, L., Manion, L., & Morrison, K. (2017). *Research methods in education* (8th ed.). Routledge.

Cole, M. (1996). *Cultural psychology: A once and future discipline*. Harvard University Press.

Cole, M. (1998). Can cultural psychology help us think about diversity? *Mind, Culture, and Activity, 5*(4), 291–304. https://doi.org/10.1207/s15327884mca0504_4

Cole, M. (2005). Using cross-cultural psychology to design afterschool educational activities in different cultural settings. In W. Friedlmeier, P. Chakkarath, & B. Schwarz (Eds.), *Culture and human development: The importance of cross-cultural research for the social sciences* (pp. 53–71). Psychology Press.

Cole, M. (2010). Cultural–historical activity theory. In *Qualitative research* (pp. 360–366). Laboratory for Comparative Human Cognition. http://lchc.ucsd.edu/People/MCole/CHAT-Paper_2010.pdf

Cole, M., & Engeström, Y. (1993). A cultural-historical approach to distributed cognition. In G. Salomon (Ed.), *Distributed cognitions: Psychological and educational considerations* (pp. 1–46). Cambridge University Press.

Council, F. A. W., & Britain, G. (1993). *Annual review*. The Council.

Craig, A. P. (1996). Education for all. *South African Journal of Higher Education, 10*(2), 47–55. https://journals.co.za/doi/pdf/10.10520/AJA10113487_1309

Creswell, J. W., & Creswell, J. D. (2018). *Research design: Qualitative, quantitative, and mixed methods approaches* (5th ed.). Sage Publications.

Daniels, H. (2002). *Vygotsky and pedagogy*. Routledge.

Daniels, H. (2005). Vygotsky and educational psychology: Some preliminary remarks. *Educational and Child Psychology, 22*(1), 6–17.

Daniels, H. (2006). The 'social' in post-Vygotskian theory. *Theory & Psychology, 16*(1), 37–49. https://doi.org/10.1177/0959354306060107

Davydov, V. V. (1975a). *Soviet studies in the psychology of learning and teaching mathematics* (Vol. 7). University of Chicago Press.

Davydov, V. V. (1975b). Logical and psychological problems of elementary mathematics as an academic subject. In L. P. Steffe (Ed.), *Soviet studies in the psychology of learning and teaching mathematics: Vol. 7. Children's capacity for learning mathematics* (pp. 55–107). School Mathematics Study Group. (Original work published 1966).

Davydov, V. V. (1982). *Soviet studies in the psychology of learning and teaching mathematics* (Vol. 7). University of Chicago Press.

de Beer, J. (2022). Looking at teacher professional development through a fourth-generation Cultural-Historical Activity Theory lens: The value of "Change Laboratories". In *Future-Proofing Teacher Education* (pp. 112–129). Routledge.

Denzin, N. K., & Lincoln, Y. S. (Eds.). (2005). *The SAGE handbook of qualitative research* (3rd ed.). Sage Publications.

Diaz, R. M., Neal, C. J., & Amaya-Williams, M. (1990). Vygotsky and education: The social origins of self-regulation. In L. C. Moll (Ed.), *Vygotsky and education: Instructional implications and applications of sociohistorical psychology* (pp. 127–154). Cambridge University Press.

Dixon, K., Janks, H., Granville, S., Kell, C., Makalela, L., & Tyler, R. (2018). A critical analysis of CAPS for Life Skills in the Foundation Phase (Grades R-3). *Journal of Education, 71*, 1–20. https://doi.org/10.17159/2520-9868/i71a01

Diyal, S. B. (2024). Integration of constructivist pedagogy and ICT tools to enhance learning outcomes. *Educational Journal, 3*(1), 59–67.

Edwards, A., & Hedegaard, M. (2023). *Taking children and young people seriously*. Cambridge University Press. https://doi.org/10.1017/9781108980869

Engeström, Y. (1987). *Learning by expanding: An activity theoretical approach to developmental research*. Orienta-Konsultit.

Engeström, Y. (1999). Activity theory and individual. *Perspectives on activity theory,* 19.

Engeström, Y. (2001). Expansive learning at work: Toward an activity theoretical reconceptualization. *Journal of Education and Work, 14*(1), 133–156. https://doi.org/10.1080/13639080123238

Engeström, Y. (2005a). *Developmental work research: Expanding activity theory in practice*. Lehmanns Media.

Engeström, Y. (2005b). *Knotworking to create collaborative intentionality capital in fluid organizational fields*.

Engeström, Y. (2008). *From teams to knots: Activity-theoretical studies of collaboration and learning at work*. Cambridge University Press.

Engeström, Y. (2009). The future of activity theory: A rough draft. In A. Sannino, H. Daniels, & K. D. Gutiérrez (Eds.), *Learning and expanding with activity theory* (pp. 303–328). Cambridge University Press. https://doi.org/10.1017/CBO9780511809989.020

Engeström, Y. (2010). Activity theory and learning at work. In M. Malloch, L. Cairns, K. Evans, & B. N. O'Connor (Eds.), *The SAGE handbook of workplace learning* (pp. 86–104). Sage Publications.

Engeström, Y. (2011). From design experiments to formative interventions. *Theory & Psychology, 21*(5), 598–628.

Engeström, Y. (2014). *Learning by expanding: An activity-theoretical approach to developmental research*. Cambridge University Press.

Engeström, Y. (2018). *Expertise in transition: Expansive learning in medical work*. Cambridge University Press.

Engeström, Y., & Sannino, A. (2010). Studies of expansive learning: Foundations, findings, and future challenges. *Educational Research Review, 5*(1), 1–24. https://doi.org/10.1016/j.edurev.2009.12.002

Engeström, Y., & Sannino, A. (2011). Discursive manifestations of contradictions in organizational change efforts: A methodological framework. *Journal of Organizational Change Management, 24*(3), 368–387. https://doi.org/10.1108/09534811111132758

Engeström, Y., & Sannino, A. (2017). Studies of expansive learning: Foundations, findings and future challenges. In H. Daniels, M. Cole, & J. V. Wertsch (Eds.), *The Cambridge companion to Vygotsky* (3rd ed., pp. 1–24). Cambridge University Press.

Engeström, Y., & Sannino, A. (2020). From mediated actions to heterogenous coalitions: Four generations of activity-theoretical studies of work and learning. *Mind, Culture, and Activity, 28*(1), 4–23. https://doi.org/10.1080/10749039.2020.1806328

Engeström, Y., & Toiviainen, H. (2010). Co-configurational design of learning instrumentalities: An activity theoretical perspective. In *Learning Across Sites* (pp. 45–64). Routledge.

Feuerbach, L. (2013). Principles of the philosophy of the future (Z. Hanfi, Trans.). In Z. Hanfi (Ed. & Trans.), *The fiery brook: Selected writings* (pp. 175–246). Verso Books. (Original work published 1843)

Flavell, J. H. (2020). *The developmental psychology of Jean Piaget*. Barakaldo Books.

Fleer, M. (2011). "Conceptual play": Foregrounding imagination and cognition during concept formation in early years education. *Contemporary Issues in Early Childhood, 12*(3), 224–240.

Fleer, M., Gomes, J., & March, S. (2014). Science learning affordances in preschool environments. *Australasian Journal of Early Childhood, 39*(1), 38–48. https://doi.org/10.1177/183693911403900106

Fodor, J. A. (1980). Methodological solipsism considered as a research strategy in cognitive psychology. *Behavioural and Brain Sciences, 3*(1), 63–73. https://doi.org/10.1017/S0140525X00001515

Foot, K. A. (2014). Cultural-historical activity theory: Exploring a theory to inform practice and research. *Journal of Human Behavior in the Social Environment, 24*(3), 329–347. https://doi.org/10.1080/10911359.2013.831011

Galleguillos, J., & Borba, M. D. C. (2018). Expansive movements in the development of mathematical modeling: analysis from an Activity Theory perspective. *ZDM, 50*(1), 129–142.

Gallimore, R. (1990). Teaching mind in society: Teaching, schooling, and literate discourse. In L. Moll (Ed.), *Vygotsky and education: Instructional implications and application of sociohistorical psychology*. Cambridge University Press. https://doi.org/10.1017/CBO9781139173674.009

Gallimore, R., & Tharp, R. (1990). Teaching mind in society: Teaching, schooling. *Vygotsky and education: Instructional implications and applications of sociohistorical psychology, 175*. Cambridge University Press.

Geertz, C. (1973). *The interpretation of cultures: Selected essays*. Basic Books.

Gibran, K. (1993). *The Prophet*. A.D. Donker.

Greenberg, J., & Moll, L. C. (1990). Creating zones of possibilities: Combining social contexts for instruction. In L. C. Moll (Ed.), *Vygotsky and education: Instructional implications and applications of sociohistorical psychology* (pp. 319–348). Cambridge University.

Gu, S., Pasqualetti, F., Cieslak, M., Telesford, Q. K., Yu, A. B., Kahn, A. E., Medaglia, J. D., Vettel, J. M., Miller, M. B., Grafton, S. T., & Bassett, D. S. (2015). Controllability of structural brain networks. *Nature Communications, 6*, 8414. https://doi.org/10.1038/ncomms9414

Hardman, J. (2008). Researching pedagogy: An activity theory approach. *Journal of Education, 45*(1), 65–95.

Hardman, J. (2015). Pedagogical variation with computers in mathematics classrooms: A Cultural Historical Activity Theory analysis. *PINS, 48*, 47–76. *ISSN* 2309-8708.

Hardman, J. (2019a). Teachers' perceptions of pedagogical change with information communication technologies in no-fee primary schools in a developing context: A cultural historical activity theory analysis. *Journal of Education & Social Policy, 6*(3), 125–137. https://doi.org/10.30845/jesp.v6n3p16

Hardman, J. (2019b). Towards a pedagogical model of teaching with ICTs for mathematics attainment in primary school: A review of studies 2008–2018. *Heliyon, 5*(5).

Hardman, J. (2022). A cultural-historical pedagogical model as a potential developmental tool in schools. *European Journal of Education and Pedagogy, 3*(3), 191–198.

Hardman, J. (2023). *A cultural-historical approach towards pedagogical transitions: Transitions in Post-Apartheid South Africa*. Bloomsbury. ISBN 978-1-3501-6470-3.

Hardman, J., & Raudzingana, M. (2021). Mathletics software and student attainment in grade 4-a cultural historical analysis. *Advances in Social Sciences Research Journal, 8*(5), 517–531.

Hedegaard, M. (1998). Situated learning and cognition: Theoretical learning and cognition. *Mind, Culture, and Activity, 5*(2), 114–126. https://doi.org/10.1207/s15327884mca0502_5

Hedegaard, M. (Ed.). (2001). *Learning in classrooms*. Aarhus University Press.

Hedegaard, M., & Chaiklin, S. (2005). *Radical-local teaching and learning: A cultural-historical approach*. Aarhus University Press.

Hegel, G. W. F. (2004). *Phenomenology of spirit* (A. V. Miller, Trans.). Oxford University Press.

hn, H. (2014). Teachers' attitudes towards Korean English in South Korea. *World Englishes, 33*(2), 195–222. https://doi.org/10.1111/weng.12081

Hoadley, U. (2017). *Pedagogy in poverty: Lessons from twenty years of curriculum reform in South Africa*. Routledge. https://doi.org/10.4324/9781315680927

Hopwood, N. (2025). Twenty-five years of change laboratories in schools: A critical and formative review. *Educational Action Research, 33*(4), 571–598.

Hundeb, A. B. (2024). Effects of GeoGebra-assisted instructional methods on students' conceptual understanding of geometry. *Cogent Education, 11*(1), 2379745. https://doi.org/10.1080/2331186X.2024.2379745

Ilyenkov, E. V. (1974). Activity and knowledge. In E. V. Ilyenkov (Ed.), *Philosophy and culture*. Politizdat. http://www.marxists.org/archive/ilyenkov/works/activity/index.htm

Johri, A., Roth, W.-M., & Olds, B. M. (2013). Representations and engineering learning [Special issue]. *Journal of Engineering Education, 102*(2), 161–164. https://doi.org/10.1002/jee.20005

Kajamaa, A., Kumpulainen, K., & Rajala, A. (2023). Finding life beyond the classroom walls: A Change Laboratory supporting expansive de-encapsulation of school. *Éducation et Didactique, 17*(2), 125–141. https://doi.org/10.4000/educationdidactique.11773

Kallaway, P. (Ed.). (2002). *The history of education under apartheid, 1948-1994: The doors of learning and culture shall be opened*. Pearson South Africa.

Kanjee, A., Chetty, M., Shozi, N., Ahmed, A., & Correa, F. (2025). *TIMSS 2023: Highlights report South Africa*. Department of Basic Education. https://www.education.gov.za/Portals/0/Documents/Reports/2024/TIMSS%202024.pdf

Kaptelinin, V. (2005). The object of activity: Making sense of the sense-maker. *Mind, Culture, and Activity, 12*(1), 4–18. https://doi.org/10.1207/s15327884mca1201_2

Karanasios, S., Riisla, K., & Simeonova, B. (2017, July 6–8). Exploring the use of contradictions in activity theory studies: An interdisciplinary review. [Conference presentation]. 33rd Egos Colloquium: The Good Organization: Aspirations, Interventions, Struggles, Copenhagen, Denmark.

Kirk, J., & Miller, M. L. (1986). *Reliability and validity in qualitative research*. Sage Publications.

Koul, R. B., Sheffield, R., & McIlvenny, L. (2021). *Teaching 21st century skills: Using STEM makerspaces*. Springer.

Kozulin, A. (1986). The concept of activity in Soviet psychology: Vygotsky, his disciples and critics. *American Psychologist, 41*(3), 264–274. https://doi.org/10.1037/0003-066X.41.3.264

Kozulin, A. (2003). Psychological tools and mediated learning. In A. Kozulin, B. Gindis, V. S. Ageyev, & S. M. Miller (Eds.), *Vygotsky's educational theory in cultural context* (pp. 15–38). Cambridge University Press.

Kuiper, E., & de Pater-Sneep, M. (2014). Student perceptions of drill-and-practice mathematics software in primary education. *Mathematics Education Research Journal, 26*(2), 215–236. https://doi.org/10.1007/s13394-013-0088-1

Lam, P. L., Ng, H. K., Tse, A. H., Lu, M., & Wong, B. Y. (2021). eLearning technology and the advancement of practical constructivist pedagogies: Illustrations from classroom observations. *Education and Information Technologies, 26*(1), 89–101.

Leeds-Hurwitz, W. (2015). The roots of intercultural communication in cultural anthropology: Excavating the past to understand the present. *China Intercultural Communication Annual, 1*, 67–84.

Lektorsky, V. A. (2009). Mediation as a means of collective activity. *Learning and Expanding with Activity Theory*, 75–87.

Lenin, V. I. (1958). *Collected works* (C. Dutt, Trans., Vol. XXXVIII). Progress. https://www.marxists.org/archive/lenin/works/cw/pdf/lenin-cw-vol-38.pdf

Lenin, V. I. (1964). *Materialism and empirio-criticism*. In *Collected works* (Vol. 14). Progress Publishers. (Original work published 1909).

Leont'ev, A. N. (1974). The problem of activity in psychology. *Soviet Psychology, 13*(2), 4–33. https://doi.org/10.2753/RPO1061-040513024

Leont'ev, A. N. (1977). The dialectical method in the psychology of memory. *Soviet Psychology, 16*(1), 53–69. https://doi.org/10.2753/RPO1061-0405160153

Leont'ev, A. N. (1978). *Activity, consciousness, and personality*. Prentice-Hall.

Leont'ev, A. A. (1981). *Psychology and the language learning process*. Pergamon.

Leont'ev, D. A. (1992). Joint activity, communication, and interaction (Toward well-grounded "pedagogy of cooperation"). *Journal of Russian & East European Psychology, 30*(2), 43–58. https://doi.org/10.2753/RPO1061-0405300243

Leontiev, A. N. (1981). *Problems of the development of the mind*. Moscow: Progress. (Original work published 1959).

Li, Y., Liu, X., & Wang, Y. (2025). The impact of Information and Communication Technology (ICT) on learning outcomes in early childhood and primary education: A meta-analysis of moderating factors. *Frontiers in Psychology, 16*, 1540169. https://doi.org/10.3389/fpsyg.2025.1540169

Li, Q., & Ma, X. (2010). A meta-analysis of the effects of computer technology on school students' mathematics learning. *Educational Psychology Review, 22*(3), 215–243.

Lim, C. P. (2002). A theoretical framework for the study of ICT in schools: A proposal. *British Journal of Educational Technology, 33*(4), 411–421. https://doi.org/10.1111/1467-8535.00273

Lukács, G. (1972). *History and class consciousness: Studies in Marxist dialectics* (R. Livingstone, Trans.). MIT Press. (Original work published 1923).

Luria, A. R. (1976). *Cognitive development: Its cultural and social foundations*. Harvard University Press.

Maistry, S. M., & Africa, I. E. (2020). Neoliberal stratification: The confounding effect of the school poverty quintile ranking system in South Africa. *South African Journal of Education, 40*(4), 1–9. https://doi.org/10.15700/saje.v40n4a1872

Mao, T.-T. (1937). On practice. In *Selected works of Mao Tse-tung* (Vol. 1, pp. 295–309). Foreign Languages Press.

Marx, K. (1867). *Capital: A critique of political economy* (B. Fowkes, Trans., Vol. 1). Penguin.

Marx, K., & Engels, F. (1967). *The German ideology: Parts I & III* (R. Pascal, Ed.). International Publishers. (Original work published 1932).

Mercer, N. (2016). Education and the social brain: Linking language, thinking, teaching and learning. *Éducation et didactique, 10*(10-2), 9–23. https://doi.org/10.4000/educationdidactique.2523

Merriam, S. B. (2009). *Qualitative research: A guide to design and implementation*. Jossey-Bass.

Miles, M. B., & Huberman, A. M. (1984). *Qualitative data analysis: A sourcebook of new methods*. Sage Publications.

Moll, L. C., & Greenberg, J. B. (1990). Creating zones of possibilities: Combining social contexts for instruction. In L. C. Moll (Ed.), *Vygotsky and education: Instructional implications and applications of sociohistorical psychology* (pp. 319–348). Cambridge University Press.

Morris, A., Hardman, J., & Jacklin, H. (2016). School science for six-year-olds: A neo-Vygotskian approach to curriculum analysis. *Journal of Education, 64*, 1–26.

Muller, J. (2014). Every picture tells a story: Epistemological access and knowledge. *Education as Change, 18*(2), 255–269. https://doi.org/10.1080/16823206.2014.932256

Newman, F., & Holzman, L. (1993). *Lev Vygotsky: Revolutionary scientist*. Routledge.

Piaget, J. (1976a). *The language and thought of the child* (M. Gabain, Trans., 3rd ed.). Routledge & Kegan Paul. (Original work published 1923).

Piaget, J. (1976b). Piaget's theory. In *Piaget and his school: A reader in developmental psychology* (pp. 11–23). Springer Berlin Heidelberg.

Piaget, J. (1977). *The development of thought: Equilibration of cognitive structures* (A. Rosin, Trans.). Viking Press. (Original work published 1975).

Rodrigues, A., Camillo, J., & Mattos, C. (2011). Cultural-historical activity theory and science education: foundational principals and potentialities. *Scientific Papers, 778*, 191–200.

Rosen, Y., & Salomon, G. (2007). The differential learning achievements of constructivist technology-intensive learning environments as compared with traditional ones: A meta-analysis. *Journal of Educational Computing Research, 36*(1), 1–14.

Roth, W.-M. (2003). *Toward an anthropology of graphing: Semiotic and activity-theoretic perspectives*. Kluwer Academic Publishers.

Roth, W. M. (2004). Activity theory and education: An introduction. *Mind, Culture, and Activity, 11*(1), 1–8. https://doi.org/10.1207/s15327884mca1101_1

Roth, W. M., & Lee, Y. J. (2007). "Vygotsky's neglected legacy": Cultural-historical activity theory. *Review of Educational Research, 77*(2), 186–232. https://doi.org/10.3102/0034654306298273

Rubtsova, O. V. (2021). Experimenting with roles in adolescence: Applying drama for constructing the zone of proximal development. *Kul'turno-istoricheskaya psikhologiya = Cultural-Historical Psychology, 17*(2), 105–113. https://doi.org/10.17759/chp.2021170210

Rubtsova, O., & Daniels, H. (2016). The concept of drama in Vygotsky's theory: Application in research. *Cultural-Historical Psychology, 12*(3), 189–207.

Sabri, N. M., & Hj Daud, S. M. (2014). Effectiveness of ICT integration in schools: A survey in Malaysian schools. *International Journal of Education and Development using ICT, 10*(2), 77–94. https://files.eric.ed.gov/fulltext/EJ1105224.pdf

Sannino, A. (2008a). Experiencing conversations: Bridging the gap between discourse and activity. *Journal for the Theory of Social Behaviour, 38*(3), 267–291. https://doi.org/10.1111/j.1468-5914.2008.00371.x

Sannino, A. (2008b). From talk to action: Experiencing interlocution in developmental interventions. *Mind, Culture, and Activity, 15*(3), 234–257. https://doi.org/10.1080/10749030802186769

Sannino, A. (2010). Teachers' talk of experiencing: Conflict, resistance and agency. *Teaching and Teacher Education, 26*(4), 838–844. https://doi.org/10.1016/j.tate.2009.10.021

Sannino, A., & Engeström, Y. (2018). Cultural-historical activity theory: Founding insights and new challenges. *Cultural-Historical Psychology, 14*(3), 43–56. https://doi.org/10.17759/chp.2018140305

Sannino, A., Engeström, Y., & Lemos, M. (2018). Formative interventions for expansive learning and transformative agency. In *Cultural-historical activity theory approaches to design-based research* (pp. 113–147). Routledge.

Seloane, O. M., et al. (2023). Developing undergraduate engineering mathematics students' conceptual and procedural knowledge of complex numbers using GeoGebra. *Pythagoras, 44*(1), a763. https://doi.org/10.4102/pythagoras.v44i1.763

Slavin, R. E., Lake, C., & Groff, C. (2009). Effective programs in middle and high school mathematics: A best-evidence synthesis. *Review of Educational Research, 79*(2), 839–911.

Smagorinsky, P. (1995). The social construction of data: Methodological problems of investigating learning in the zone of proximal development. *Review of Educational Research, 65*(3), 191–212. https://doi.org/10.2307/1170682

Smagorinsky, P. (2011a). *Vygotsky and literacy research: A methodological framework*. Sense.

Smagorinsky, P. (2011b). Vygotsky's stage theory: The psychology of art and the actor under the direction of *perezhivanie*. *Mind, Culture, and Activity, 18*(4), 319–341. https://doi.org/10.1080/10749039.2010.518300

Smagorinsky, P. (2018). Deconflating the ZPD and instructional scaffolding: Retranslating and reconceiving the zone of proximal development as the zone of next development. *Learning, culture and social interaction, 16*, 70–75. https://doi.org/10.1016/j.lcsi.2017.10.009

Spinuzzi, C. (2020). "Trying to predict the future": third-generation activity theory's codesign orientation. *Mind, Culture, and Activity, 27*(1), 4–18.

Stake, R. E. (1995). *The art of case study research*. Sage Publications.

Stake, R. E. (2006). *Multiple case study analysis*. Guilford Press.

Stetsenko, A. (2005). Activity as object-related: Resolving the dichotomy of individual and collective planes of activity. *Mind, Culture, and Activity, 12*(1), 70–88. https://doi.org/10.1207/s15327884mca1201_6

Stetsenko, A. (2021). Scholarship in the context of a historic socioeconomic and political turmoil: Reassessing and taking stock of CHAT. Commentary on Y. Engeström and A. Sannino "from mediated actions to heterogenous coalitions: four generations of activity-theoretical studies of work and learning". *Mind, Culture, and Activity, 28*(1), 32–43. https://doi.org/10.1080/10749039.2021.1874419

Street, B. V. (Ed.). (2001). *Literacy and development: Ethnographic perspectives*. Routledge.

Tamim, R. M., Bernard, R. M., Borokhovski, E., Abrami, P. C., & Schmid, R. F. (2011). What forty years of research says about the impact of technology on learning: A second-order meta-analysis and validation study. *Review of Educational research, 81*(1), 4–28.

Thanh, N. C., & Thanh, T. T. L. (2015). The interconnection between interpretivist paradigm and qualitative methods in education. *American Journal of Educational Science, 1*(2), 24–27.

Tharp, R. G., & Gallimore, R. (1988). *Rousing minds to life: Teaching, learning, and schooling in social context*. Cambridge University Press.

Tomaz, V. S., & David, M. M. (2021). Transfer of learning as boundary crossing between cultural-historical activity systems. In *Transfer of learning: Progressive perspectives for mathematics education and related fields* (pp. 221–249). Springer International Publishing.

Uden, L. (2007). Activity theory for designing mobile learning. *International Journal of Mobile Learning and Organisation, 1*(1), 81–102.

Vasilyuk, F. E. (1988). *The psychology of experiencing: An analysis of how critical situations are dealt with* (R. English, Trans.). Progress Publishers. (Original work published 1984).

Veresov, N. (2015). Duality of categories or dialectical concepts? *Integrative Psychological and Behavioral Science, 50*(2), 244–256. https://doi.org/10.1007/s12124-015-9327-1

Virkkunen, J., & Newnham, D. S. (2013). *The Change Laboratory: A tool for collaborative development of work and education*. Sense Publishers.

Virkkunen, J., Mäkinen, E., & Lintula, L. (2013). From diagnosis to clients: Constructing the object of collaborative development between physiotherapy educators and workplaces. In H. Daniels, A. Edwards, Y. Engeström, T. Gallagher, & S. R. Ludvigsen (Eds.), *Activity theory in practice: Promoting learning across boundaries and agencies* (pp. 25–40). Routledge.

Vygotsky, L. S. (1978a). Interaction between learning and development. *Readings on the Development of Children, 23*(3), 34–41.

Vygotsky, L. S. (1978b). *Mind in society: The development of higher psychological processes* (Vol. 86). Harvard University Press.

Vygotsky, L. S. (1979). Consciousness as a problem in the psychology of behavior. *Soviet Psychology, 17*(4), 3–35. https://doi.org/10.2753/RPO1061-0405170403

Vygotsky, L. S. (1986). *Thought and language* (E. Hanfmann & G. Vakar, Trans.). MIT Press.

Vygotsky, L. S. (1987). *The collected works of L. S. Vygotsky: The fundamentals of defectology* (Vol. 2). Springer Science & Business Media.

Vygotsky, L. S. (1993). The fundamentals of defectology (Abnormal psychology and learning disabilities) (J. E. Knox & C. B. Stevens, Trans.). In R. W. Rieber & A. S. Carton (Eds.), *The collected works of L. S. Vygotsky: Vol. 2*. Plenum Press. (Original work published 1983).

Vygotsky, L. S. (1994). The problem of the environment. In R. van der Veer & J. Valsiner (Eds.), *The Vygotsky reader* (pp. 338–354). Blackwell.

Vygotsky, L. S. (1997). *The collected works of L. S. Vygotsky: Problems of the theory and history of psychology* (R. W. Rieber & J. Wollock, Eds., Vol. 3). Springer. https://doi.org/10.1007/978-1-4615-5893-4

Vygotsky, L. S. (1998). *The collected works of L. S. Vygotsky: Vol. 5. Child psychology* (R. W. Rieber, Ed.; M. J. Hall, Trans.). Plenum Press.

Vygotsky, L. S. (2012). *Thought and language* (Vol. 29). MIT Press.

Wegerif, R. (2011). From dialectic to dialogic. In T. Koschmann (Ed.), *Theories of learning and studies of instructional practice* (pp. 201–221). Springer.

Wells, G. (1999). *Dialogic inquiry* (pp. 137–141). Cambridge Press.

Wertsch, J. V. (1990). Dialogue and dialogism in a socio-cultural approach to mind. In I. Marková & K. Foppa (Eds.), *The dynamics of dialogue* (pp. 62–82). Harvester Wheatsheaf.

Wertsch, J. V. (1991a). A sociocultural approach to socially shared cognition. In L. B. Resnick, J. M. Levine, & S. D. Teasley (Eds.), *Perspectives on socially shared cognition* (pp. 85–100). American Psychological Association. https://doi.org/10.1037/10096-004

Wertsch, J. V. (1991b). *Voices of the mind: Sociocultural approach to mediated action*. Harvard University Press.

Wertsch, J. V. (1998). *Mind as action*. Oxford University Press.

Wertsch, J. V. (2003). Commentary on: Deliberation with computers: Exploring the distinctive contribution of new technologies to collaborative thinking and learning. *International Journal of Educational Research, 39*(8), 899–904. https://doi.org/10.1016/j.ijer.2004.11.012

Wertsch, J. V. (2007). National narratives and the conservative nature of collective memory. *Neohelicon, 34*(2), 23–33. https://doi.org/10.1007/s11059-007-2003-9

White, C. J., & Van Dyk, H. (2019). Theory and practice of the quintile ranking of schools in South Africa: A financial management perspective. *South African Journal of Education, 39*(Supplement 1), s1–19. https://doi.org/10.15700/saje.v39ns1a1820

Wood, D., Bruner, J. S., & Ross, G. (1976). The role of tutoring in problem solving. *Journal of Child Psychology and Psychiatry, 17*(2), 89–100. https://doi.org/10.1111/j.1469-7610.1976.tb00381.x

World Bank. (2024). *Government expenditure on education, total (% of GDP)* [Data set]. https://data.worldbank.org/indicator/SE.XPD.TOTL.GD.ZS

Yamagata-Lynch, L. C. (2003). Using activity theory as an analytic lens for examining technology professional development in schools. *Mind, Culture, and Activity, 10*(2), 100–119. https://doi.org/10.1207/S1532-7884MCA1002_2

Yamagata-Lynch, L. C. (2010). *Activity systems analysis methods: Understanding complex learning environments*. Springer.

Yin, R. K. (1984). *Case study research: Design and methods*. Sage Publications.

Yin, R. K. (2009). *Case study research: Design and methods* (4th ed.). Sage Publications.

The manufacturer's authorised representative in the EU is Springer Nature Customer Service Centre GmbH, Europaplatz 3, 69115 Heidelberg, Germany. If you have any concerns regarding our products, please contact ProductSafety@springernature.com

Printed and bound by CPI Group (UK) Ltd, Croydon, CR0 4YY
20/10/2026
02236474-0008